Astrochemistry

Chemistry in Interstellar and Circumstellar Space

Astrochemistry

Chemistry in Interstellar and Circumstellar Space

By

David A. Williams
University College London, UK
Email: daw@star.ucl.ac.uk

and

Cesare Cecchi-Pestellini
Osservatorio Astronomico di Palermo, Italy
Email: cesare.cecchipestellini@inaf.it

ROYAL SOCIETY
OF **CHEMISTRY**

Print ISBN: 978-1-83916-396-8
EPUB ISBN: 978-1-83916-939-7

A catalogue record for this book is available from the British Library

The Royal Society of Chemistry is a charity, registered in England and Wales, Number 207890, and a company incorporated in England by Royal Charter (Registered No. RC000524), registered office: Burlington House, Piccadilly, London W1J 0BA, UK, Telephone: +44 (0) 20 7437 8656.

Visit our website at www.rsc.org/books

Printed in the United Kingdom by CPI Group (UK) Ltd, Croydon, CR0 4YY, UK

Preface

Astrochemistry is now a well-established subject. For chemists, astrochemistry provides an opportunity to study how chemistry may operate in extreme and hostile environments; it makes huge demands on both laboratory and theoretical chemistry and consequently stimulates new activities in these studies. For astronomers, astrochemistry has revealed previously unknown molecular components of our galaxy, the Milky Way, and of external galaxies; these newly-recognised components play a fundamental role in the evolution of galaxies and have revolutionized our understanding of the Milky Way and of our place in it.

The aim of this book is to introduce some of these exciting ideas to students of chemistry. We describe the processes by which gas phase chemistry occurs in interstellar and circumstellar gases. This chemistry creates specific molecular tracers of those regions, triggers the growth of solid particles allowing surface and solid state chemistry to occur, and generates a remarkably rich range of molecular species which may even have links to astrobiology. The presence of the most abundant molecular species, such as molecular hydrogen and carbon monoxide, influences the evolution of interstellar gas, permitting the formation of dense cores of gas in which the formation of stars, planets, comets and asteroids may occur.

We cannot hope to cover all topics of relevance to astrochemistry in this introductory text. For example, we have not discussed at appropriate length the roles of gas dynamics, the thermal balance in interstellar gas, or the range of timescales that should be considered, although these are important topics. Our intention is to concentrate

Astrochemistry: Chemistry in Interstellar and Circumstellar Space
By David A. Williams and Cesare Cecchi-Pestellini
© David A. Williams and Cesare Cecchi-Pestellini 2023
Published by the Royal Society of Chemistry, www.rsc.org

on the chemical aspects of the subject rather than the physical; current texts generally emphasize the latter. We hope that this book will provide interest and encouragement for some readers to take the subject to a higher level. The structure of the book is described below.

Chapter 1 introduces the astronomical background, while Chapter 2 discusses some (probably familiar) aspects of the interaction of molecules with electromagnetic radiation. In Chapter 3 we describe the types of reactions that are most relevant to astrochemistry, while in Chapter 4 the difficulties of initiating interstellar chemistry are described. The challenges are overcome by the introduction of highly specific initiating chemical pathways. It becomes clear at this point that the reaction network necessary for a comprehensive description of interstellar chemistry can be very large; typically, these networks involve reactions between hundreds of species interacting in thousands of reactions. Fortunately, large libraries of reactions (with measured, computed, or estimated rate coefficients) are available, together with many computer programmes that enable the chemical rate equations to be integrated. Chapter 5 introduces these topics and shows how, with suitable computer programmes, a student may make rapid progress in following the evolution of chemistry in a variety of astronomical environments. Thus, Chapter 5 is an important chapter that draws together all the ideas presented in the book up to this point and should be able to give the student reader an impressive skill in studying astrochemistry.

As mentioned in Chapter 1, interstellar dust is present in all interstellar gas. Dust scatters and absorbs starlight so that the interiors of gas clouds are shielded from stellar radiation (as described in Chapter 5) and molecules survive more easily in those locations. Where and how is dust formed? These are the topics of Chapter 6, which describes how the seemingly unlikely situations in the atmospheres of cool stars and in the apparently even more unlikely envelopes of supernovae ejecta, fairly simple gas phase chemistry can ultimately lead to the growth of dust particles. These particles (or grains) are then ejected into the interstellar medium where they may evolve further. Dust grains do have active chemical roles, beyond the rather passive role of shielding clouds from stellar radiation. These active roles include providing surfaces on which chemistry may occur (described in Chapter 7) and on which mixed molecular ices may in some situations accumulate (Chapter 8). These ices are deposited in rather simple chemical forms, but – as demonstrated in many laboratory experiments – when suitably energized by fast particles or by energetic radiation provide an efficient chemistry leading to the formation of a wide range of relatively large organic molecules, many of which are species that are astronomically detected. Evidently, solid state chemistry plays a key role in astrochemistry. Chapter 9 describes how planet formation is a by-product of star formation and how large organic molecules may be transported onto a planet. We discuss the boundaries between astrochemistry and astrobiology, and the possible relevance that molecules such as amino acids formed in ices may have in abiogenesis. Finally, Chapter 10 looks back to questions raised in Chapter 1, to discuss the extent to which those questions have been answered in the course of this book.

We hope that the user of this book will experience the thrills and enjoyment in learning about a field of study that is interdisciplinary and open-ended, feelings that we ourselves share when thinking about astrochemistry.

David A. Williams
University College London
Cesare Cecchi-Pestellini
INAF – Observatory of Palermo

Units in Astronomy

Units of distance used in this book are the *astronomical unit* (AU), the *parsec* (pc), and the *light-year* (ly). The **astronomical unit** is defined to be the mean Earth-Sun distance, and so

$1\ \text{AU} = 1.496 \times 10^8\ \text{km}$

The **parsec** is defined in terms of the astronomical unit. It is the distance at which the Earth-Sun distance of 1 AU subtends one second of arc (*i.e.*, 1/3600 of one degree), so that

$1\ \text{pc} = 206\,265\ \text{AU} = 3.086 \times 10^{13}\ \text{km}$

The **light-year** is defined as the *distance* travelled at the speed of light in a vacuum ($2.9979 \times 10^8\ \text{m s}^{-1}$) in one year (365.25 days, or $3.1558 \times 10^7\ \text{s}$), so that

$1\ \text{ly} = 9.461 \times 10^{15}\ \text{m}$

and therefore

$1\ \text{pc} = 3.26\ \text{ly}$

The **unit of astronomical mass** is usually taken as the solar mass, $M_\odot$, determined to be

$1\ M_\odot = 1.989 \times 10^{30}\ \text{kg}$

In these units, for example, the Milky Way galaxy (which is the main focus of our attention in this book) has a mass of about $10^{12}\ M_\odot$, a diameter of about 27 kpc (kiloparsec), and the distance of the Sun from the galactic centre is about 8 kpc. Thus, distances within the Milky Way galaxy are typically measured in kpc. Distances between the Milky Way and external galaxies are typically measured in units of Mpc.

Astrochemistry: Chemistry in Interstellar and Circumstellar Space
By David A. Williams and Cesare Cecchi-Pestellini
© David A. Williams and Cesare Cecchi-Pestellini 2023
Published by the Royal Society of Chemistry, www.rsc.org

Number Densities and Rate Coefficients in Astrochemistry

Basic ideas about astrochemistry began to be explored theoretically in the middle of the 20th century when it was conventional to use centimetre-gram-seconds (cgs) units rather than SI. This convention became embedded in the astrochemical literature (a few examples of which are appended to each chapter of this book) and research papers in astronomy are conventionally written using a mixture of cgs and the units of astronomy, as appropriate. Therefore, we have retained this convention in this book, so that our readers may consult the astrochemical literature more easily.

Number densities are given per cm^3 rather than per m^3, and rate coefficients of binary reactions are in units of $cm^3\ s^{-1}$ rather than m^3 s^{-1}. Thus, a binary rate coefficient of (say) $10^{-17}\ m^3\ s^{-1}$ is replaced by one of $10^{-11}\ cm^3\ s^{-1}$. We like to think that this convention has some small but particular value in astronomy, because the mean number density of hydrogen atoms in the interstellar medium of the Milky Way galaxy is about one per cm^3 (or $10^6\ m^{-3}$), and so all other number densities (say, of molecular clouds with a hydrogen number density of $10^4\ cm^{-3}$) are automatically compared to this mean.

Astrochemistry: Chemistry in Interstellar and Circumstellar Space
By David A. Williams and Cesare Cecchi-Pestellini
© David A. Williams and Cesare Cecchi-Pestellini 2023
Published by the Royal Society of Chemistry, www.rsc.org

Contents

Astrochemistry: Chemistry in Interstellar and Circumstellar Space
By David A. Williams and Cesare Cecchi-Pestellini
© David A. Williams and Cesare Cecchi-Pestellini 2023
Published by the Royal Society of Chemistry, www.rsc.org

3 Gas Phase Reactions in Interstellar and Circumstellar Media   49

4 Gas Phase Chemical Networks in Interstellar Clouds   72

7 Surface Chemistry on Interstellar Dust Grains   158

8 Interstellar Ices and Solid-state Chemistry as a Route to Molecular Complexity   171

9 Interstellar Chemistry, Astrobiology, and the Origin of Life   185

10 Conclusions   227

Subject Index   242

1 What is Astrochemistry?

"To date, 241 individual molecular species, comprising 19 different elements, have been detected in the interstellar and circumstellar medium by astronomical observations. These molecules range in size from two atoms to seventy, and have been detected across the electromagnetic spectrum from centimetre wavelengths to the ultraviolet." (Brett A. McGuire)[1]

1.1 Astronomical Background

1.1.1 A Brief Historical Introduction

Astrochemistry is a fairly new and quite rapidly developing interdisciplinary subject, so it seems appropriate to begin with a brief account of how the subject began. Up to the 1930s, it was known that clouds of gas existed in the space between the stars in our galaxy, the Milky Way. Could the gas be partially molecular? It wasn't known. Careful theoretical studies, notably by the world famous astronomer, Sir Arthur Eddington, in 1926, concluded that no known processes in this gas could produce molecules at a sufficient rate that detectable abundances of molecules could be created.[2]

However, just over a decade later, in 1937, astronomers discovered for the first time that the optical signatures of a few simple molecular species were present in spectroscopic observations of the very low

Astrochemistry: Chemistry in Interstellar and Circumstellar Space
By David A. Williams and Cesare Cecchi-Pestellini
© David A. Williams and Cesare Cecchi-Pestellini 2023
Published by the Royal Society of Chemistry, www.rsc.org

density gas occupying the vast space between the stars in the Milky Way galaxy.[3,4] The first molecular species to be detected in interstellar gas were CH, CH^+, and CN, and their optical lines were seen in absorption along lines of sight towards several bright stars which acted as continuum sources of radiation. The molecular abundances were deduced to be very low. This first discovery of molecules in space was considered at the time to be interesting, but merely a backwater of contemporary astronomy, which at the time was mainly concerned with studying the evolution of stars. In fact, we now know that the first detections of these three interstellar molecular species were just the first steps in the discovery of a rich astronomical chemistry and in the development of what we now call *astrochemistry*. This new branch of science has provided a revolution in the way that astronomers understand the evolution of the Milky Way and all galaxies, and has brought new understanding of the formation of stars and planets: in fact, the role of molecules in astronomy is fundamental. At the time, however, it was recognised that – just as Eddington had found – the formation of molecules in the very low density interstellar gas where they are subjected to intense ultraviolet radiation from massive stars presented challenging chemical problems to which no convincing solutions could be offered. These and other problems are now largely resolved, and the special chemistry that has been discovered to operate under the physical conditions that astronomy provides is the main concern of this book.

In the years following the early molecular detections, astronomers began to use radio telescopes to explore the galaxy. The discovery in 1951 of interstellar emission from *atomic* hydrogen in the famous 21 cm line enabled maps of the Milky Way to be made, revealing its spiral structure. It became clear that the interstellar medium is almost entirely hydrogen, with helium being present with an abundance by number of atoms of about a tenth that of hydrogen; the main reactive elements are present merely in trace amounts.

A quarter-century after the first detections of the interstellar molecular species CH, CH^+, and CN, another molecular species, the hydroxyl radical, OH, was detected (in 1963) in interstellar space.[5] This detection was made using the new waveband for molecular detections: radio. The detection signature was absorption by foreground OH molecules of background radio emission, in a hyperfine transition in the molecule corresponding to a radio wavelength of 18 cm. This molecule was also detected in intense emission lines from material around an exploding star, a supernova. The emission intensity was found to be so great that it was attributed to stimulated emission: this was the first maser known in astronomy. Interstellar ammonia, NH_3, (in 1968) and water, H_2O, (in 1969) were also detected at radio wavelengths.

Another new waveband technique for molecular astronomy soon became available: ultraviolet spectroscopy carried out using rocket-borne spectrometers above the Earth's atmosphere (which is opaque in the far-ultraviolet). Later, ultraviolet spectrometers were carried on orbiting satellites. Rocket-borne observations of interstellar gas on lines of sight towards bright stars revealed for the first time in 1970 the presence of ultraviolet absorption lines of *molecular* hydrogen.[6] Together with the earlier discovery of interstellar atomic hydrogen by its emission at 21 cm, these observations of molecular hydrogen showed that a significant fraction of the total amount of interstellar hydrogen could be in molecular form. Later studies showed that on lines of sight where molecular hydrogen was abundant, other interstellar molecules were also more likely to be found. This association between the relatively abundant molecular hydrogen and other trace species suggested that molecular hydrogen played a fundamental role in interstellar chemistry.

Soon after the detection of interstellar molecular hydrogen, a flood of detections of new interstellar molecular species began. It continues to this day. Very many of these detections were made at millimetre and sub-millimetre wavelengths where molecular emissions or absorptions in rotational transitions often arise. One of the most important and abundant interstellar molecular species is carbon monoxide, CO. It was first detected in 1970, by its emission at a wavelength of 2.6 mm,[7] corresponding to the lowest energy and longest wavelength rotational transition of CO. This molecule is, after molecular hydrogen, the most abundant interstellar molecular species, and it has been used widely to trace the presence of interstellar gas and to show the structure of molecular gas in the Milky Way and other galaxies. CO traces interstellar gas that is denser than that in the low density clouds in which CH, CH^+, and CN were first detected. These CO observations mapped a population of cold, relatively dense, dark molecular clouds throughout the Milky Way galaxy; these clouds had been discovered earlier by their ammonia emissions.

The list of interstellar molecular species detected by their spectra over the electromagnetic spectrum from radio to far-ultraviolet has grown very significantly over the last half-century. It includes many molecular species that seem familiar from school chemistry (such as water, ammonia, methyl alcohol, *etc.*) but also many less familiar species. A few of these less familiar species were actually discovered to be present in interstellar space even *before* they had been synthesized and studied in the laboratory (*e.g.*, cyanotetraacetylene, HC_9N). Evidently, a rich and unusual chemistry is occurring in the interstellar gas in the Milky Way and other galaxies, in spite of what seem to be (see Table 1.3b, below) quite unfavourable physical conditions for an active chemistry. We'll discuss the chemistry that generates *interstellar molecules* in Chapters 3 and 4.

This astronomical chemistry is not limited to clouds of gas in interstellar space, far from stars. Molecules are also found in regions close to some types of stars. These circumstellar regions may have physical conditions, such as higher densities and warmer temperatures, which may be quite favourable to the operation of an effective chemistry. However, some special circumstellar regions appear to be quite hostile to chemistry (such as circumstellar gas around supernova explosions) but may be – rather surprisingly – fairly rich in molecules. We'll discuss in Chapter 6 the chemistries that produce these *circumstellar* molecules.

If molecular species found in circumstellar regions are included in the list of detected interstellar species, then the total number of all interstellar and circumstellar species is currently about 250. We'll show and discuss a complete (up to mid-2021) table of detected interstellar and circumstellar species in Section 1.3 of this chapter. If detected isotopologues (*i.e.*, molecules that differ only in their isotopic composition, such as $^{12}C^{16}O$, $^{12}C^{17}O$, $^{12}C^{18}O$, $^{13}C^{16}O$, $^{13}C^{17}O$, and $^{13}C^{18}O$) are also included, then the total number of detected species is probably several times larger than 250. The actual number of interstellar molecular species (detected and undetected) must be even larger; our studies in Chapter 4 will imply that many more species must be involved in the chemical networks that form the detected molecules, and these species must also be present. These additional species may be undetected because their abundance is too low for detection or they may be undetectable if they have no transitions in suitable regions of the electromagnetic spectrum.

For the astronomer, the molecules provide excellent probes of the physical conditions in interstellar and circumstellar regions. For the chemist, these regions provide an opportunity to explore gas phase, surface, and solid-state chemistry operating under physical conditions that are very different from those normally found on Earth and which are generally much more hostile to chemistry. To understand the chemistries that are functioning in space requires inputs from across a wide range of science: from astronomical observers who provide the fundamental data on molecular abundances and the physical conditions where the molecules are found, from computational modellers studying the networks of chemical processes that may give rise to the observed species, from astrophysicists who describe the remarkable role of molecules in the evolution of galaxies, and from laboratory and theoretical chemists who measure or compute essential data on reaction rate coefficients that are required in those computations. The interaction and cooperation between all these disciplines has created the new subject of astrochemistry. A gentle introduction to astrochemistry can be found in ref. 8.

In this book, we shall explore the chemistry that provides the rich array of molecular species that are found in interstellar and circumstellar regions, and describe the roles that these molecules play in the formation of stars and planets.

1.1.2   The Milky Way

The most detailed and complete observational information about astrochemistry is, of course, for our own galaxy, the Milky Way, so we'll summarize its main features first. But we note here that the Milky Way is merely one of very many galaxies that make up the observable Universe. We'll discuss very briefly in Section 1.1.5 the variety of galaxy types observed in the Universe. To a greater or lesser extent, depending on the physical conditions, the processes of astrochemistry that occur in the Milky Way also occur in external galaxies.

Our planet Earth orbits our star, the Sun, which is merely one of a huge number of stars in the Milky Way galaxy. In terms of its mass and intrinsic brightness, the Sun is a fairly typical member of this collection of stars; a few stars are much more massive and very much brighter, but many stars are fairly similar to the sun or less massive (down to about one tenth of a solar mass) and consequently fainter. There are believed to be a few times 10^{11} stars in the Milky Way and they are distributed approximately in a disc shape, roughly circular, with a spiral structure within the disc and a bar structure through the centre. The Sun is not located in the galactic Milky Way in a privileged position, but is located about one half of a radius from the galactic centre, and in the central plane of the disc. As seen from Earth on a moon-free night, the distribution of stars in the galaxy appears as a dense band of stars, many of which are too faint for the naked eye to distinguish. We see the light of unresolved stars as a diffuse (or milky) haziness, hence the name: the Milky Way.

The dimensions of the Milky Way are huge: the stellar disc's diameter is about 170 000 light years (roughly 50 kiloparsecs, or kpc; see tables in the preface for information on units used in astronomy) and the thickness of the stellar disc is about 0.6 kpc. On average, therefore, stars are separated by a few parsecs from each other, but in fact the distribution of stars in the galaxy is very far from uniform. The mass of the galaxy resides mainly in the stars, and is estimated to be about 10^{12} solar masses (or $M_\odot$). The whole structure, the barred spiral, is rotating. Some information about the Milky Way is summarized in Table 1.1. Of course, we can't obtain an image showing the large scale structure of the entire Milky Way as would be seen from outside the galaxy, since we are embedded within it. But we can examine images of galaxies that seem to be very similar in size and structure, and two such examples are shown in Figure 1.1. These images reveal the spiral

Table 1.1 Some approximate data for the Milky Way galaxy.

Diameter	27 kpc
Thickness of stellar disc	0.6 kpc
Number of stars	10^{11}
Total mass	$10^{12}\,M_\odot$
Interstellar mass	$10^{11}\,M_\odot$
Sun's distance from centre	8 kpc
Spiral pattern rotation period	$3 \times 10^8\,y$
Galaxy type	Barred spiral

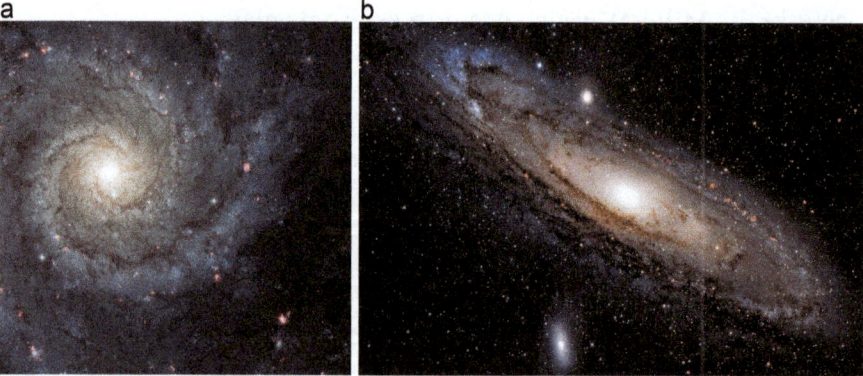

a   b

Figure 1.1 Images of two spiral galaxies believed to be similar to the Milky Way. (a) shows an optical image of galaxy M74 which we see nearly face-on (Credit: NASA, ESA, and the Hubble Heritage (STSC/AURA)-ESA/Hubble Collaboration; Acknowledgement: R. Chandler (University of Toledo) and J. Miller (University of Michigan)), and (b) shows the Andromeda galaxy which we see nearly edge-on. The spiral structure is evident in both images, and Andromeda shows a very thin disc. Reproduced from https://commons.wikimedia.org/w/index.php?curid=12654493 under the terms of the CC BY 2.0 license http://creativecommons.org/licenses/by/2.0/deed.en.

structure and the concentration of matter into a very thin disc in these galaxies. Both these features are present in the Milky Way.

We shall discuss in Chapter 6 the chemistry that occurs in cool stellar envelopes of stars of modest mass and also in the supernovae explosions that end the lives of more massive stars. These near-stellar regions are much denser than interstellar molecular clouds, and the temperatures within them are also much higher than the generally low temperatures of interstellar molecular clouds. It is not surprising, therefore, that a rich chemistry occurs so that these near-stellar regions can be largely molecular. Some of these stars eject gas and dust into interstellar space during a brief late stage of their evolution. Supernovae explosions from the relatively few stars of high mass (from about ten up to several tens of solar masses) also expel gas and dust into interstellar space. They also have a very significant large scale effect on gas in the galaxy, creating a very hot ($\sim 10^6$ K), very low density ($\sim 10^{-3}$ cm^{-3}) ionized gas that fills much of interstellar space (see Table 1.3a, below).

1.1.3   Interstellar Gas in the Milky Way

The space between the stars of the Milky Way is occupied by a very tenuous and dusty gas: *the interstellar medium*. This gas is almost entirely hydrogen, with traces of other elements (see Table 1.2). The mass of the interstellar medium is about one percent of the mass of the galaxy. If this large mass were spread over the entire volume of the galaxy it would amount to an average number density of about one hydrogen atom per cubic centimetre, an exceptionally low number density. As we can see from the elemental abundances in Table 1.2, these very low number densities would preclude the possibility of chemistry occurring on a reasonable timescale in the galaxy. Fortunately, the interstellar medium is far from uniform. The interstellar medium contains a variety of structures, with hydrogen number densities ranging from

Table 1.2   Approximate abundances by number, relative to hydrogen, of the most abundant elements in the Milky Way galaxy.

Element	Abundance	Element	Abundance
H	1	Si	3.16×10^{-5}
He	9.77×10^{-2}	Mg	3.63×10^{-5}
O	5.75×10^{-4}	Fe	3.31×10^{-5}
C	2.14×10^{-4}	S	1×10^{-5}
N	6.2×10^{-5}	Na, Ca	2×10^{-6}

Table 1.3 (a) Hot and (b) cool regions of the interstellar medium.

(a) Hot regions

Name	Filling factor in interstellar space	Typical dimension in interstellar space (pc)	Number density of H^+ ions (cm^{-3})	Temperature (K)
Coronal gas	~50%	1000	10^{-3}	10^6
Warm ionized gas	~40%	1000	0.3	8000
HII regions	<1%	1	10^2–10^4	8000
Warm neutral gas	~10%	100	0.3	8000

(b) Cool regions

Name	Typical size (pc)	Number density, H (cm^{-3})	Temperature (K)	State of hydrogen	Chemistry	Ice mantles on dust?	Dynamical age (My)
Cool neutral	10	30	100	H	No	No	10
Diffuse	10	10^2	100	H/H_2	Yes	No	3
Translucent	1	10^3	30	H_2/H	Yes	No	3
Molecular	0.1	10^4	10	H_2	Yes	Yes	1
IRDC[a]	0.1	10^5	25	H_2	Yes	Yes	0.01
"Hot" core	0.01	10^7	300	H_2	Yes	No	0.01

[a]Infrared dark cloud.

about $\sim 10^{-3}$ cm^{-3} to $\sim 10^{7}$ cm^{-3} (or even higher). Interstellar molecules are generally located in fairly cool interstellar gas with densities above about 10^{2} cm^{-3}. The gas in interstellar clouds is not uniform in density, but has a complex filamentary structure. Bright spots along these filaments are associated with the arrival of new stars forming from the interstellar gas.

The Milky Way (and probably the Universe, too) is flooded with *cosmic rays*. These "rays" are, in fact, fast-moving particles; they are atomic nuclei, and most of them are hydrogen nuclei (protons). They are probably accelerated to high velocities by extreme shocks arising from supernovae and active galactic nuclei. Energies for a single particle up to or larger than 10^{20} eV have been inferred. One of these very high energy atomic particles has kinetic energy comparable to that of an apple falling under Earth's gravity through several metres! In fact, cosmic rays with these extreme energies are very rare, and the much more numerous cosmic ray particles with much lower energies (about a few MeV per particle) are responsible for most of the influence of these particles on the gas. Cosmic rays ionize interstellar atoms and molecules, ejecting energetic electrons into the gas. The energy is shared through collisions, and so this amounts to an important heating effect in the interstellar medium. This heating may help to generate structures (such as clouds) in the interstellar gas. The ionization of molecules by cosmic rays creates molecular ions; however, these ions are short-lived (as we'll discuss in Chapter 3), and so cosmic ray ionization tends to suppress molecular abundances, affecting the cooling of molecular gas, its temperature and pressure, and hence its structure.

The electromagnetic radiation in the interstellar gas includes, of course, starlight from the many stars in the galaxy, especially in the optical and ultraviolet parts of the spectrum. Some of this radiation is absorbed by dust grains that are everywhere mixed with interstellar gas (we shall discuss the origin and properties of dust grains in Section 1.1.4, and also in the later chapters of this book) and re-radiated at longer wavelengths, so the Milky Way dust is also a source of infrared radiation. The Galaxy is also a source of radio waves emitted from ions moving through the tangled magnetic field lines that thread the Galaxy. Powerful events both inside the Galaxy and external to it generate X-rays and γ-rays. Thus, photons of the complete range of the electromagnetic spectrum pervade the interstellar medium and have a significant effect on the interstellar gas on both the local and large-scale interstellar structures. Some of these photon-driven processes are discussed in more detail in Chapter 3.

Table 1.3(a) and (b) summarize the approximate characteristics of the various types of interstellar region identified in the Milky Way. These structures are maintained by the impact of strong radiation fields, both particle radiation and electromagnetic radiation, on the interstellar gas. The numbers included in these tables are approximate and indicative rather than specific.

The data in Table 1.3(a) show that nearly all of the interstellar volume is occupied by gas that is either at very high temperatures ($\sim 10^6$ K) and very low density ($\sim 10^{-3}$ cm^{-3}) or warm (~ 8000 K) and low density (~ 0.3 cm^{-3}). These structures are broadly maintained by heating caused by supernovae explosions. These gases are fully ionized. The very hot gas is called "coronal" gas because ultraviolet absorption lines from ionized oxygen, carbon, and nitrogen arise in this gas, and these lines were originally identified in the solar corona. Recombination in the warm ionized gas generates a warm neutral low density gas. The components called HII regions are the regions surrounding hot stars and ionized by ultraviolet starlight. These regions are roughly spherical, and can be seen in optical photographs (such as M51 shown in Figure 1.2) as regions of red emission at a wavelength of 656 nm corresponding to the Balmer transition 3→2 in the recombination spectrum of H$^+$ with electrons.

None of these "hot" components is capable of generating a significant chemistry as molecules are readily destroyed under these conditions. Their importance in the present context is that these "hot" components occupy nearly all of interstellar space but contain only a tiny fraction of interstellar mass; therefore, the remaining gas – which comprises nearly all of the interstellar mass – is confined to very small

Figure 1.2   Image of spiral galaxy NGC 2403, showing strings of near-spherical HII regions along the spiral arms. (Credit: Suburu Telescope (NAOJ), Hubble Legacy archive, processing by Robert Gendler.)

regions of space. This contrast may be illustrated by comparing the characteristic sizes of "hot" and "cold" regions in Tables 1.3(a) and (b).

Table 1.3(b) shows the types of interstellar region that have been observationally identified and which are relatively cool and dense. Note that the numerical values shown are merely indicative of typical rather than precise values. The *cool neutral regions* are of low density; they are pervaded by stellar ultraviolet radiation which dominates any chemistry occurring. However, the remaining regions are those in which molecules may be readily observed and an active chemistry is operating. *Diffuse* clouds arc the type of cloud in which the first three detections of interstellar molecular species were made in 1937. The number of species detected in such clouds is now much greater and the chemistry is evidently fairly complex even in these rather low density clouds. There is some protection for the chemical processes against destruction by ultraviolet starlight provided by some extinction by interstellar dust. These diffuse clouds probably survive for millions of years until dynamical events associated with supernovae destroy them. *Translucent clouds* benefit from higher densies and increased extinction, and represent a transition region between diffuse clouds and molecular clouds.

Molecular clouds are denser regions that are almost entirely molecular. In them, ultraviolet radiation from nearby massive stars is largely excluded because of extinction caused by interstellar dust (see sub-section 1.1.4). The chemistry in molecular clouds is mainly determined by cosmic ray ionization. These molecular clouds may be dense enough to be gravitationally unstable, and if so they may not survive for more than a million years. In the process of collapse to higher densities, chemistry continues and becomes more complex. Ices are deposited on the surfaces of dust grains, and solid-state chemistry within these ices generates new molecular species. Regions like this are believed to be the sites where massive stars are formed; such regions are called *infrared dark clouds* (IRDCs). Eventually, the warming of the central regions by radiation from the protostar (the newly forming star) releases complex molecules from the ice mantles on dust grains. These late stages in the evolution have relatively short lifetimes, much less than a million years, and are known as *hot cores*.

1.1.4 Interstellar Dust in the Milky Way

We've referred several times already to the roles that interstellar dust may play in astrochemistry, and the catalytic activity of dust will be described in detail in the second part of this book. However, it will be useful in this rapid survey of the interstellar medium in the Milky Way to summarize in this section some of the main information about the origin and properties of interstellar dust.

The suggestion that there might be some material in interstellar space tending to obscure the light of background stars was made by William Herschel, who in 1784 was using a telescope to make simple sketches of rich star fields. He noticed in one such star field a small zone in which there appeared to be a complete absence of stars, and famously noted (in German) in his journal "Here indeed is a hole in the Heavens!" He may have believed that such a "hole" truly existed, or he may have considered that some localized material in front of the star field extinguished the light of the background stars, giving the appearance of a "hole". Later, more sensitive photographic techniques showed that some stars not detected by Herschel did exist in the putative "hole", confirming that the more likely explanation was of some obscuring material in the line of sight, rather than of a true absence of stars. Further observations showed that the obscuration was not just a local effect but a general phenomenon: there is a widespread interstellar "fog" – called *interstellar extinction* – along almost all lines of sight in the Galaxy (and, indeed, in external galaxies, too). This extinction causes a partial dimming of starlight everywhere; sometimes the dimming can be effectively total, just as a thick terrestrial fog can extinguish the light from a nearby streetlight. An example of extinction of a portion of a rich star field is shown in Figure 1.3.

Further, it was found that the amount of extinction caused by the interstellar fog varied with frequency in such a way that starlight in the blue part of the spectrum was more heavily extinguished than red light. Thus, distant stars were apparently "reddened" compared to closer similar stars. This was a useful phenomenon, allowing estimates of distances to stars to be made, since more reddened stars should be located at greater distances. Theoretical studies showed that this colour-dependent (*i.e.*, wavelength-dependent) behaviour in the visual spectrum could be accounted for if there was a population of spherical dust grains with a range of diameters comparable to the wavelengths of visible light, *i.e.*, around some hundreds of nanometres in diameter. Assuming that the amount of dust is proportional to the amount of gas along a line of sight, the fogginess can in principle be used to measure the amount of interstellar gas along that particular line of sight.

When studies of extinction were able to be extended into the infrared and ultraviolet, it was clear that extinction in the infrared continued to decrease as wavelength increased, whereas extinction in the

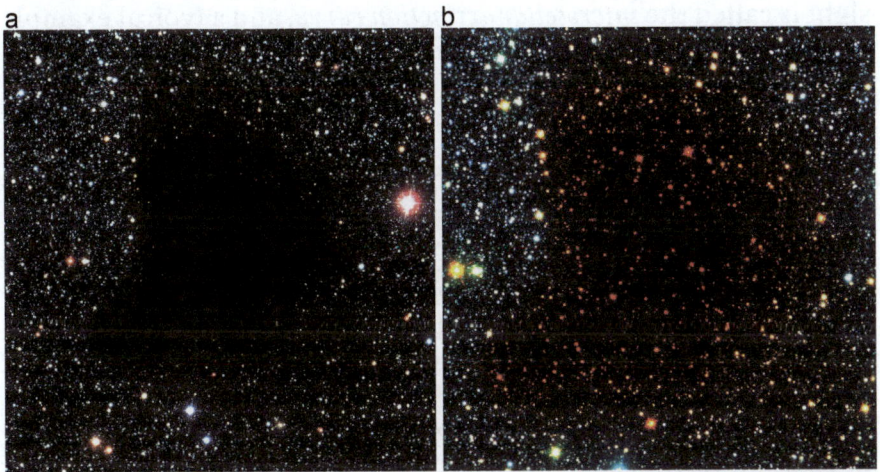

Figure 1.3 A rich star field photographed in the optical (a) and also in the infrared (b). An intervening dusty gas cloud extinguishes the optical light of the background stars (a), but the infrared image (b) reveals their presence. (Credit for both images: ESO.)

ultraviolet was more complicated. There is a localised peak in extinction in the near ultraviolet at a wavelength of about 220 nm, and then after a decline to a local minimum at a wavelength near 160 nm there is a strong rise in extinction in the far ultraviolet extinction (~100 nm). Astronomers measure extinction in magnitudes, a term reaching back to ancient astronomy when the brightest stars were said to be of first magnitude, the next brightest were said to be of second magnitude, *etc.* When numerical studies could eventually be made, it was found that a first magnitude star was about one hundred times brighter than a sixth magnitude star. This relationship could be expressed mathematically in the following way:

$$A_\lambda = 2.5 \log_{10}(I_0/I_\lambda) \tag{1.1}$$

so a change in intensity caused by extinction from I_0 to I_λ by a factor of 100 corresponds to a change in extinction, A_λ, of 5 magnitudes. The variation of interstellar extinction with wavelength along a particular line of sight is generally measured relative to the extinction at a particular wavelength in the visual region of the spectrum, denoted by V; this reference wavelength is conventionally chosen to be 550 nm. The curve that shows how this normalized extinction varies with wavelength along a particular line of sight is called the *interstellar extinction curve*, and a typical example of a normalized interstellar extinction curve in the Milky Way is shown in Figure 1.4. This figure shows how the extinction caused by interstellar dust varies with wavelength along a typical line of sight in the Milky Way.

In Table 1.3(b), we list the cool regions of the interstellar medium; these are the regions which have the potential to be partly or largely molecular. Since the gas and dust are relatively well mixed, the density gives a measure of the extinction in a region. The cool neutral regions have little extinction and are freely pervaded by the mean interstellar radiation field. Diffuse clouds have a modest amount of extinction, up to about one magnitude of extinction in the visible, and more extinction in the ultraviolet (according to the curve shown in Figure 1.5). Translucent clouds may have about three magnitudes of visual extinction. The three remaining types, molecular clouds, the IRDCs, and Hot Cores, are denser and have so much extinction that one may assume that the external interstellar radiation field is almost entirely excluded by the extinction caused by interstellar dust.

On sufficiently long paths through the Milky Way galaxy, the amount of dust increases linearly with path length, at a rate of about 1.8 visual magnitudes per kiloparsec. On the large scale, assuming that dust and gas are well mixed, it is found that the average column density, N_H, of total hydrogen in both atoms H and molecules, H_2, and the average

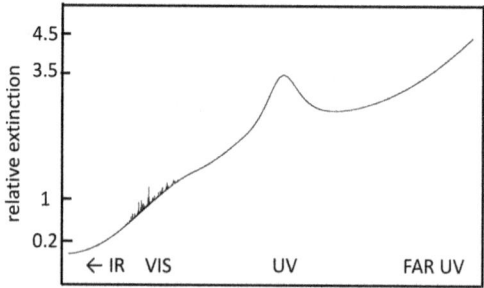

Figure 1.4   Characteristic interstellar extinction curve in the Milky Way. This diagram shows how the average extinction measured in the Milky Way galaxy varies with wavelength. In the infrared, extinction is small compared to that in the visible. Extinction increases almost linearly with inverse wavelength in the visible, and rises to a peak near 220 nm in the near ultraviolet. Further into the ultraviolet, extinction first decreases then rises strongly in the far ultraviolet, ~100 nm, where the amount of extinction is about three times that in the visible.

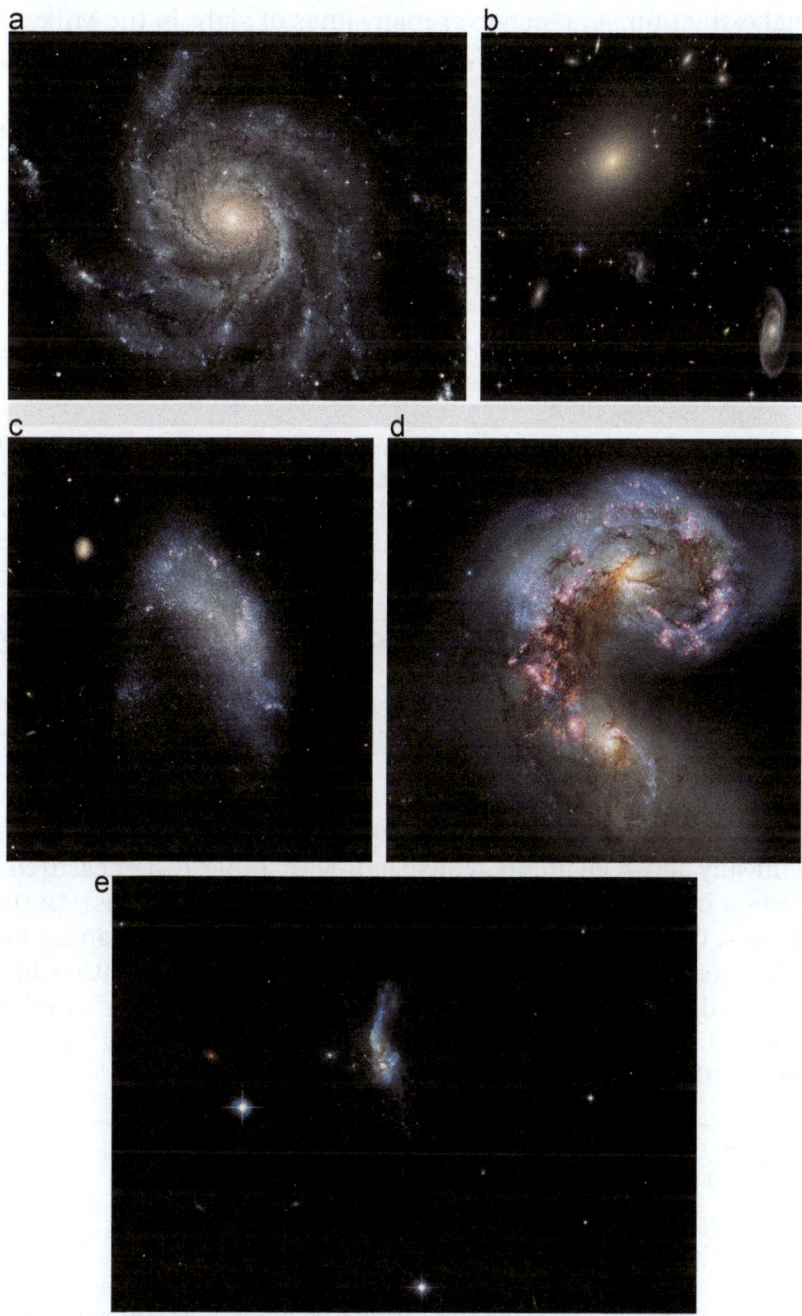

Figure 1.5 Some examples of images of various types of galaxy. (a) The Pinwheel galaxy (also called M101) is a spiral galaxy and from Earth is viewed face-on. (Credit: NASA, ESA, and the Hubble Space Telescope.) (b) The elliptical galaxy ESO-325-G004 has a

(continued)

visual extinction, A_V, taken over many lines of sight in the Milky Way are related in the following equation:

$$N_H/A_V = 1.9 \times 10^{21} \text{ cm}^2 \text{ per magnitude} \qquad (1.2)$$

The idea that interstellar extinction is caused by a population of small particles in interstellar space is supported by other lines of evidence and is now accepted. This evidence includes

(i) the collection and identification of dust grains within the solar system; some of the collected grains clearly originate from outside the solar system;

(ii) detected interstellar absorption features in the infrared attributed to absorption in solid state material;

(iii) detected scattering and polarization of starlight; both of these phenomena are attributed to the interaction of starlight with solid particles; for polarization, at least some of the dust grains must be asymmetric and weakly aligned, probably by a magnetic field;

(iv) detected thermal emission (in the infrared) from solid particles heated by starlight;

(v) the observed absence (or the so-called *depletion*) of some gas phase atoms where dust grains are abundant. See Table 1.4 for information about interstellar depletions.

Table 1.4 shows the total relative abundances of the most abundant chemically active elements (consistent with Table 1.2), measured for hot stars in the Milky Way (relative to 10^6 hydrogen atoms). In these hot stars, the temperatures are so great that dust grains cannot exist. In the interstellar medium, however, elements may be either in the gas or in dust grains. The third column shows the measured relative elemental abundances in the diffuse interstellar gas. The difference between the second and third columns (the "missing" material)

Figure 1.5 smooth profile; none of the stars in this galaxy are resolved, and *(continued)* most of the stars are of relatively low mass. (Credit: NASA, ESA, and the Hubble Heritage Team and J. Blakeslee.) (c) Galaxy NGC 1427A is an irregular galaxy. These galaxies are small and their shapes are affected strongly by near-collisions with massive galaxies. (Credit: NASA, ESA, and the Hubble Heritage Team.) (d) The Antennae galaxies are galaxies (NGC 4038 and 4038) in collision, generating a high star formation rate. Such galaxies are called starburst galaxies. (Credit: ESA, Hubble, and NASA.) (e) This galaxy, IRAS 14348-1447, is an example of an Ultraluminous Infrared Galaxy (ULIRG). ULIRGs have very high star formation rates. (Credit: ESA, Hubble and NASA.)

Table 1.4 Interstellar depletions of elements in diffuse interstellar gas.

Element	Abundance	Interstellar gas	Interstellar dust
O	575	389	186
C	214	91	123
N	62	62	0
Mg	36.3	1.5	34.8
Fe	33.1	0.3	32.8
Si	31.6	2.2	29.4

shown in the fourth column is assumed to be in interstellar dust grains in diffuse interstellar gas.

The implication of Table 1.4 is that about one third of oxygen, more than half of carbon, and almost all of magnesium, iron and silicon are in dust grains, while all nitrogen is in the gas phase. The results suggest that the composition of interstellar dust could largely be made up of iron/magnesium silicates and carbons in solid form and in large hydrocarbons such as polycyclic aromatic hydrocarbons (PAHs). These conclusions are consistent with the results of modelling interstellar extinction curves. The models typically require a range of silicate dust grain radii from a few nm to a few hundred nm, for which the size distribution is very strongly skewed to the smallest sizes (typically, in these models, the number of grains with radii in the range a to $(a + da)$ is proportional to $a^{-3.5}da$. In such a distribution, there would be roughly three thousand times as many grains of radius 30 nm as 300 nm). A significant amount of carbon is required to be in interstellar PAHs, but solid interstellar carbon is also necessary, either as discrete grains or as a coating on silicate grains.

The conclusions of the previous paragraph depend on results of detailed calculations that describe the interaction of electromagnetic radiation with a sphere of specified refractive index. The calculation involves the application of Maxwell's equations of electromagnetism to a grain, and is simple in concept but complicated in practice. The calculation gives information about the scattering and absorption of radiation by the grain. For simplicity, let's consider the extinction caused by a population of spherical dust grains of radius a, all of the same material; the number density of the dust grains is n_d and the path length is l. Then, the intensity I_0 is reduced to

$$I_\lambda = I_0 \exp[-n_d \pi a^2 Q_{ext(\lambda)} l] = I_0 \exp(-\tau_\lambda) \qquad (1.3)$$

where Q_{ext} is an efficiency factor for extinction, and is the main result from the calculation. Here, τ_v is the optical depth and equal to

$A_v/1.086$ (where A_v is in magnitudes). In fact, the extinction is a result of two separate processes: *scattering* of radiation out of the path by the dust grains, and *absorption* of radiation by the dust grains. The efficiencies of these two processes are Q_{sca} and Q_{abs}, respectively, and

$$Q_{ext} = Q_{sca} + Q_{abs} \qquad (1.4)$$

The ratio $\omega = Q_{sca}/Q_{ext}$ measures the ability of the grains to reflect radiation, rather than absorb it, and is called the *albedo*. The calculation of these efficiencies is beyond the scope of this book. In general, the results show a near linear behaviour for $x = 2\pi a/\lambda$ on the order of unity, and approaching a constant value of 2 for large values of this same parameter, x. It is the linear behaviour that provided support for the idea that extinction is caused by dust grains, because it matches the near-linear behaviour of the interstellar extinction curve (Figure 1.4) in the visual part of the spectrum. This part of the interstellar extinction curve therefore requires the presence of some dust grains with sizes comparable to the wavelength of visual light. Similarly, the rise in extinction observed in the far ultraviolet suggests that dust grains are present with sizes much smaller than the wavelength of visible light.

We shall discuss some aspects of the formation of dust grains later in this book (see Chapter 6), but for now we shall simply note that the grains are formed (similar to the ways in which smoking candles create soot, or incomplete combustion in internal combustion engines makes sooty exhaust fumes) in cool stellar envelopes and in energetic stellar explosions, and ejected into interstellar space. These grains are not immutable and may change their sizes and physical and chemical nature after they leave the circumstellar environment and make their journey through interstellar space. They pervade the entire interstellar medium and are generally mixed uniformly with the interstellar gas. The total mass of dust in the Milky Way is about one percent of the mass of gas in the interstellar space of the Milky Way, or about 10^9 $M_\odot$. However, the dust-to-gas ratio may vary significantly from one galaxy to another, depending on stellar activity and the survivability of the dust. Although the dust is apparently a minor component of matter in the Milky Way, it has important consequences for interstellar chemistry. For our purposes in this book, the importance of dust lies mainly in its various roles in interstellar chemistry, and we shall discuss these in Chapters 7–9. The information that can be obtained about interstellar dust and its roles in chemistry are summarized in Table 1.5. A simple introduction to the properties and roles of interstellar dust can be found in ref. 9.

Table 1.5 A summary of information about interstellar dust and its roles in the Milky Way galaxy.

Origin	Circumstellar envelopes of cool, evolved stars
	Ejecta from supernovae
Signature	Extinction and polarization of starlight
Inferences from modelling interstellar extinction	Caused by small dielectric particles
	Typical range of grain radii: $a \sim 5$ nm–0.5 μm
	Typical size distribution: number in range $a \rightarrow a + da$ is $dn \sim a^{-3.5}da$
Composition	Silicate and carbon solids, either distinct or combined; usually amorphous, plus PAH molecules
Role in interstellar chemistry	Surface reactions, especially H_2 formation
Role in solid-state chemistry	Accumulation of simple mixed ices on surfaces of dust particles
	Processing of mixed ices to more complex species
Role in evolution of the galaxy	Essential functions in star and planet formation

1.1.5 Galaxies Outside the Milky Way

The observable Universe is believed to contain a very large number of galaxies, possibly as many as 10^{12} galaxies, distributed throughout the Universe in a volume of radius around 15 gigaparsec (Gpc). On average, the separation between galaxies is typically measured in units of Gpc, very much greater than the dimensions of galaxies themselves (the Milky Way diameter being about 27 kpc). The galaxies are observed to be receding from each other, with velocities increasing (and apparently accelerating) with separation. Intergalactic space cannot be completely empty since galaxies are continually ejecting matter; however, no intergalactic gas has been detected so far.

Galaxies aren't all like the Milky Way and other spiral galaxies (discussed above, see subsections 1.1.2 and 1.1.3), but are found in various shapes and sizes, and the interstellar gas within them may have a variety of physical conditions, different from those of the Milky Way. Other types of galaxies include elliptical galaxies, starburst galaxies, and irregular galaxies.

Ellipticals have, as the name implies, an ellipsoidal shape, seen from any angle. The stars are in orbit about the common centre of mass but may not be resolved. Ellipticals have few high mass stars and little structure, generally appearing quite smooth. They usually have little interstellar gas; and the abundance of heavy elements is low so that a rich chemistry is unlikely in these objects. Giant ellipticals also form, and these are much larger than typical spirals.

Some galaxies are observed to be forming stars at a very high rate. They have large reserves of interstellar gas and are converting this gas rapidly into stars, so star formation in these galaxies – the so-called *starburst galaxies* – is intense. These galaxies are often associated with galaxies that are merging together. Mergers may also give rise to LIRGs (*luminous infrared galaxies*) and ULIRGs (*ultraluminous infrared galaxies*). These objects contain large amounts of gas and dust. They have a high star formation rate, but most of the optical and ultraviolet starlight is absorbed by the dust and re-emitted in the infrared.

Active galaxies are those galaxies that have an active nucleus. The source of radiation from these galaxies is dominated by the energy released as matter is accreted on to a supermassive black hole located at the centre of the galaxy. Very energetic processes at the centre of the disc carry matter away from the disc in two opposed high speed jets.

Many galaxies don't show the symmetry found in spirals or ellipticals, and are called *irregular galaxies*. They are often small, and the irregular shape may be caused by a near collision with a larger galaxy.

Table 1.6 lists characteristic data for some galaxy types and Figure 1.5 shows examples of these galaxy types.

1.2   Evolution in the Interstellar Medium

All the data presented for the Milky Way galaxy so far in this chapter are "snapshots": they represent the state in which we see the galaxy at present. However, it seems that everything in the Milky Way – and in other galaxies, too – is evolving. The data in Table 1.3 strongly suggest an evolving picture in which low density interstellar gas is swept up by supernova shock waves to form denser neutral gas clouds. These clouds develop an internal chemistry that provides molecules; by radiating energy away from the cloud, these molecules (particularly CO molecules) help to cool the cloud and maintain a low temperature, and the clouds become mainly molecular, and denser and colder. Gravity takes a stronger and stronger hold of the cloud – or on parts of it – and the gravitational collapse of a portion of the cloud results

Table 1.6   Data for some galaxy types.

	Description	Intrinsic luminosity, $L_\odot$	Mass range, $M_\odot$	Diameter, kpc
Spiral	Flat disc, spiral arms	10^8–10^{11}	10^9–10^{12}	10–100
Elliptical	Smooth ellipsoidal	10^6–10^9	10^6–10^{13}	1–200
Irregular	Irregular, dusty	10^6–10^9	10^6–10^{11}	1–10
Starburst	Intense star formation	10^9–10^{11}	10^6–10^{10}	10^2–10^3
ULIRG	Most emission in IR, very bright	$>10^9$	10^9 (?)	10^3 (?)

eventually in the formation of structures such as infrared dark clouds in which star formation may occur. The even denser gas around a newly-formed star (a hot core) is warmed by the radiation of that star, and the molecules released in the warming are a signature of the presence of the still deeply-embedded (and optically obscured) young star.

If sufficiently massive, a newly-formed star will – after a relatively short life of a few million years (in which it generates an *HII region*) – end its life as a supernova. The supernova explosion disrupts the star-forming cloud and sends shock waves into the surrounding interstellar medium, generating the very high temperature "*coronal*" gas that fills so much of space (see Table 1.3). The supernova also seeds the interstellar with the "ashes" of the thermonuclear burning that powered the star (these ashes consist mainly of the elements shown in Table 1.2) and of dust formed in the ejecta from the supernova. Less massive stars that are formed in the collapse of a molecular cloud evolve more slowly than more massive stars. Near the end of their lives, they pass through a phase in which they too eject the ashes of their nuclear burning into interstellar space (much more gently than in a supernova explosion), together with dust formed in their fairly cool envelopes. The roles of stars as sources of gas and dust for the interstellar medium will be discussed further in Chapter 6.

The very hot "*coronal*" gas takes a long time to cool, but eventually the gas reaches a lower temperature (the *warm ionized* gas) and begins to recombine (to form the *warm neutral* gas). Eventually, the cycle begins again: low density, mainly neutral, interstellar gas accumulates because of interstellar gas dynamical events and gravity, and ultimately forms new stars. Interstellar chemistry and dust both play crucial roles in this cycle of events. Together, chemistry and dust provide molecules that control the cooling processes in interstellar gas and allow gravity to take control. Dust is, of course, essential for planet formation (as we shall see in Chapter 9).

A schematic diagram of the evolution in a galaxy like the Milky Way is shown in Figure 1.6.

We can see from this model of galaxy evolution that the variety of galaxies that is observed isn't simply random. There are reasons why one galaxy differs from another. For example, the abundances of elements and of interstellar dust depend on the cumulative amount of star formation in a galaxy; star formation in a galaxy depends on gas density and temperature. These must obviously be crucial parameters in determining interstellar chemistry. As we'll see in Chapter 4, the flux of cosmic rays drives chemistry in dark clouds, and supernovae not only inject elements into interstellar space, they also are responsible

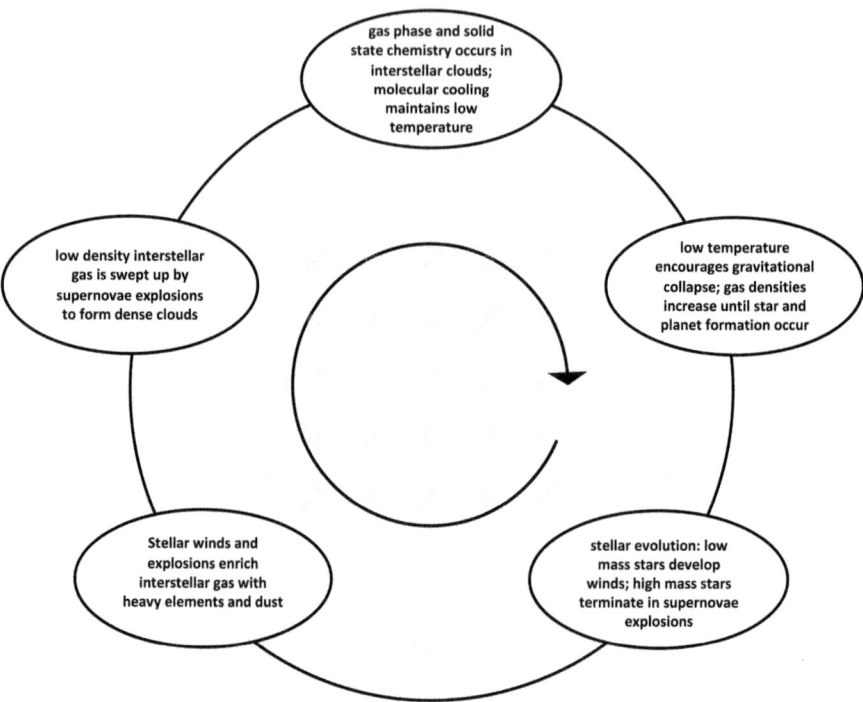

Figure 1.6 Evolution in a galaxy. The diagram indicates evolution in a galaxy like the Milky Way. Chemistry occurs in a molecular cloud,
and radiation emitted by the molecules and dust maintains a low
temperature, allowing gravity to dominate over pressure so the
cloud collapses. Star and planet formation occurs. At the ends
of their lives, low mass stars may develop winds while high mass
stars may explode in supernovae. Winds and explosions enrich
interstellar gas with heavy elements and dust. Low density interstellar gas is swept up by supernovae explosions into denser
clouds, and the cycle repeats.

for generating cosmic rays. So, galaxies that are richer in massive stars
should be richer in elements with greater atomic mass than hydrogen
and also richer in interstellar dust. These galaxies should therefore be
richer in interstellar molecules, and the effect of these molecules is
to drive more star formation, generating more supernovae, and more
cosmic rays, and so on – until the amount of interstellar gas used up
in star formation begins significantly to deplete a galaxy of interstellar
gas. Then the star formation rate declines. In general, spiral galaxies
are rich in interstellar gas and have relatively high star formation rates
(the star formation rate in the Milky Way is about one star per year for
the entire galaxy, a substantial rate; but some spiral galaxies may have
an even higher rate). Ellipticals, by contrast, have little interstellar

gas, and a low star formation rate. On the other hand, starburst galaxies have a very high rate of star formation generated in the collisions of galaxies and the rapid compression of interstellar gas. ULIRGs have even higher star formation rates than starburst galaxies.

1.3 Detected Interstellar and Circumstellar Species

Now that we have set the astronomical scene and stated the appropriate physical parameters for various regions of interstellar and circumstellar space, we can focus more specifically on the chemistry occurring in these regions. One way to do this is to examine the list of detected molecular species and to consider the difficulties faced with understanding how this variety of molecules might be produced in the interstellar medium and in circumstellar regions including ejecta from stellar explosions. The list of new detections in interstellar and circumstellar sources, shown in Appendix, grows each year, and the discovery rate is very high at present. The list shown below is believed to be correct and complete as of mid-2021 and includes more than 250 species.

Many of the data contained in the table are taken from a recent very comprehensive review[1] and a constantly updated complete list may be found in ref. 10. Both sources give a reference to the discovery paper, for each of the species listed. Note that astronomers refer to all these species as molecules, whereas many of them should strictly be referred to as radicals. However, because of the tenuous conditions in interstellar and circumstellar space, radicals have a long lifetime in space and so may be reasonably classified as molecules. Under terrestrial conditions, these radical species could only have a fleeting existence.

Each entry in the table shows the chemical symbol, the conventional name, the date of detection, and the type of region (as defined in subsection 1.1.3) in which the molecular species exists, where *cse* denotes circumstellar envelope, *dc* denotes diffuse cloud, *mc* denotes molecular cloud, *sfr* denotes star forming region, and *snr* denotes supernova remnant. For each group of molecules (diatomics, triatomics, *etc.*), the molecules are listed in order of their date of discovery by astronomers. For brevity, isotopologues have not been included in the table (apart from one case); however, it should be noted that they are common in the interstellar medium – and often highly overabundant compared to the relative abundance of the main isotopologue.

The list is impressive in the number of species and in the range of the chemistry shown. Molecular species are found in a wide variety of objects not only in the Milky Way but also outside it, in external galaxies. Given the huge intergalactic distances and the consequent difficulties of making detections, it is remarkable that as many as one third of all known species detected in the Milky Way have also been identified in one or more external galaxies. Inside the Milky Way, molecular species are found not only in the gas phase but also in the solid state (*i.e.*, in molecular ices). Molecules are also detected in exotic regions such as the atmospheres of exoplanets, but planetary detections are not included in this table. As far as the relatively dense cold gas in the Milky Way is concerned (comprising almost all of the interstellar mass), we seem to be living in a largely chemical galaxy.

A few questions about the listed species may immediately come to mind, such as:

1. Galaxies are sources of powerful particle and electromagnetic radiations that are hostile to molecules, so can we devise chemical networks that successfully overcome these losses and produce species in the abundances observed?
2. Much of interstellar gas is at very low densities and temperatures, and chemistry is likely to be slow, so is there enough time available for molecular abundances to grow?
3. The interstellar gas is H-rich and other elements are present in trace values, so why are some (but not all) of the observed species H-poor (*i.e.*, chemically unsaturated)?
4. The list of detected species includes relatively large numbers of diatomic and triatomic species, and ends (with very few exceptions) with relatively small numbers of molecular species containing about a dozen atoms. Is this decline a real chemical effect or is it associated with observational difficulties detecting the larger species? Does the number and variety of interstellar species really decline with complexity?
5. Is the chemistry producing the fullerenes – and possibly the PAHs – simply an extension of the chemistry producing the simpler molecules listed in the appendix to this chapter, or are different processes likely to be required to form these more complex molecules?
6. Silicon and iron have quite similar relative elemental abundances in the interstellar medium of the Milky Way, so why does Si appear in 13 detected species while Fe appears in only one?
7. Why are some detected interstellar molecular species known to be in the solid state? How are they formed? Do these molecules ever appear in the gas phase?

8. Dust has a passive involvement in interstellar chemistry by locking up some elements that would otherwise be present in the gas. What are its active roles? How important are they in the evolution of a galaxy like the Milky Way? Are some environments in the Universe dust-free?

9. The list includes carbon species in the form of both rings and chains. Some recent detections are of PAH molecules and one 2021 detection is of indene (fused benzene and cyclopentene rings), while the chain $HC_{11}N$ (cyanopenta-acetylene) was also detected in 2021. Is it likely that both types of structure – rings and chains – are formed in similar ways?

10. In Section 1.1.1, we noted that isotopologues are present and may be relatively abundant in space. How are these isotopologues formed? Why are isotopologues involving minor isotopes often much more abundant than expected?

Many similar questions could be devised. Obviously, the major tasks are to identify the appropriate chemical mechanisms and to devise chemical networks that might operate under the physical conditions described earlier in this chapter, in the interstellar medium of the Milky Way and other galaxies. We may then see if they can provide suitable answers to these and other questions. That is the purpose of this book. We'll return to these issues in the final chapter of this book.

Appendix: Detected interstellar and circumstellar molecular species

Diatomic molecules

CH (methylidyne) 1937, dc; **CN** (cyanide) 1937, dc; **CH$^+$** (methylidyne cation) 1937, dc; **OH** (hydroxyl) 1963, snr; **CO** (carbon monoxide) 1970, sfr; **H$_2$** (molecular hydrogen) 1970, dc; **SiO** (silicon monoxide) 1971, sfr; **CS** (carbon monosulfide) 1971, sfr; **SO** (sulfur monoxide) 1973, sfr; **SiS** (silicon monosulfide) 1975, cse; **NS** (nitrogen sulfide) 1975, sfr; **C$_2$** (diatomic carbon) 1977, sfr; **NO** (nitric oxide) 1978, sfr; **HCl** (hydrogen chloride) 1985, sfr; **NaCl** (sodium chloride) 1987, cse; **AlCl** (aluminium monochloride) 1987, cse; **KCl** (potassium chloride) 1987, cse; **AlF** (aluminium monofluoride) 1987, cse; **PN** (phosphorus mononitride) 1987, sfr; **SiC** (silicon carbide) 1989, cse; **CP** (carbon monophosphide) 1989, cse; **NH** (nitrogen monohydride) 1991, dc; **SiN** (silicon mononitride) 1992, cse; **SO$^+$** (sulfur monoxide cation) 1992, mc; **CO$^+$** (carbon monoxide cation) 1993, cse; **HF** (hydrogen fluoride) 1997, mc; **LiH**

(lithium hydride)? 1998, mc; **FeO** (ferrous oxide) 2002, sfr; **N$_2$** (molecular nitrogen) 2004, mc; **CF$^+$** (fluoromethylidynium cation) 2006, sfr; **PO** (phosphorus monoxide) 2007, cse; **AlO** (aluminium monoxide) 2009, cse; **CN$^-$** (cyanogen anion) 2010, cse; **OH$^+$** (hydroxyl cation) 2010, sfr; **SH$^+$** (sulfur monohydride cation) 2010, sfr; **O$_2$** (molecular oxygen) 2011, dc; **HCl$^+$** (hydrogen chloride cation) 2012, sfr; **SH** (sulfur monohydride) 2012, cse; **TiO** (titanium oxide) 2013, cse; **ArH$^+$** (argonium cation) 2014, snr; **CrO** (chromium monoxide) 2015, cse; **NS$^+$** (nitrogen sulfide cation) 2018, mc, sfr; **HeH$^+$** (helium hydride ion) 2019, cse; **VO** (vanadium oxide) 2019, cse;

Triatomic molecules

H$_2$O (water) 1969, sfr; **HCO$^+$** (formylium cation) 1970, sfr; **HCN** (hydrogen cyanide) 1971, sfr; **OCS** (carbonyl sulfide) 1971, sfr; **HNC** (hydrogen isocyanide) 1972, sfr; **H$_2$S** (hydrogen sulfide) 1972, sfr; **N$_2$H$^+$** (protonated nitrogen) 1974, sfr; **C$_2$H** (ethynyl) 1974, sfr; **SO$_2$** (sulfur dioxide) 1975, sfr; **HCO** (formyl) 1976, sfr; **HNO** (nitroxyl) 1977, sfr; **OCN$^-$** (cyanate) 1979, cs; **HCS$^+$** (thioformyl cation) 1981, sfr; **HOC$^+$** (hydroxymethyliumidene cation) 1983, sfr; **SiC$_2$** (cylacyclopropynylidene) 1984, cs; **C$_2$S** (dicarbon sulfide) 1987, mc, cs; **C$_3$** (tricarbon)1988, cs; **CO$_2$** (carbon dioxide) 1989, mc; **CH$_2$** (methylene) 1989, sfr; **C$_2$O** (dicarbon monoxide) 1991, mc; **MgNC** (magnesium isocyanide) 1993, cs; **NH$_2$** (amidogen) 1993, sfr; **NaCN** (sodium cyanide) 1994, cs; **N$_2$O** (nitrous oxide) 1994, sfr; **MgCN** (magnesium cyanide) 1995, cs; **H$_3$$^+$** (protonated molecular hydrogen) 1996, sfr; **SiCN** (silicon monocyanide) 2000, cs; **AlNC** (aluminium isocyanide) 2002, cs; **SiNC** (silicon mono-isocyanide) 2004, cs; **HCP**(phospha-ethyne) 2007, cs; **CCP** (dicarbon phosphide) 2008, cs; **AlOH** (aluminium hydroxide) 2010, cs; **H$_2$O$^+$** (water cation) 2010, sfr; **H$_2$Cl$^+$** (chloronium cation) 2010, sfr; **KCN** (potassium cyanide) 2010, cs; **FeCN** (iron cyanide) 2011, cs; **HO$_2$** (hydroperoxyl) 2012, mc; **TiO$_2$** (titanium dioxide) 2013, cs; **CCN** (cyanomethylidine) 2014, cs; **Si$_2$C** (disilicon carbide) 2015, cs; **S$_2$H** (hydrogen disulfide) 2017, mc; **HCS** (thioformyl) 2018, mc; **HSC** (isothioformyl) 2018, mc; **NCO** (isocyanate) 2018, mc; **CaNC** (calcium isocyanide) 2019, cse; **NCS** (thiocyanogen) 2021, sfr;

Molecules containing 4 atoms

NH$_3$ (ammonia) 1968, sfr; **H$_2$CO** (formaldehyde) 1969, sfr; **HNCO** (isocyanic acid) 1972, sfr; **H$_2$CS** (thioformaldehyde) 1973, sfr; **C$_2$H$_2$** (acetylene) 1976, cs; **C$_3$N** (cyanoethynyl) 1977, cs; **HNCS** (isothiocyanic acid)

1979, sfr; **HOCO⁺** (protonated carbon dioxide) 1981, sfr; **C₃O** (tricarbon monoxide) 1984, mc; *l*-**C₃H** (propynylidyne) 1985, mc, cs; **HCNH⁺** (protonated hydrogen cyanide) 1986, sfr; **H₃O⁺** (protonated water) 1986, sfr; **C₃S** (tricarbon monosulfide) 1987, mc; *c*-**C₃H** (cyclopropenylidyne) 1987, mc, cs; **HC₂N** (cyanocarbene) 1991, cs; **H₂CN** (methylene amidogen) 1991, cs; **SiC₃** (silicon tricarbide) 1999, cs; **CH₃** (methyl) 2000, sfr; **C₃N⁻** (cyanoethynyl anion) 2008, cs; **PH₃** (phosphine) 2008, cs; **HCNO** (fulminic acid) 2009, mc; **HOCN** (cyanic acid) 2009, sfr; **HSCN** (thiocyanic acid) 2009, sfr; **HOOH** (hydrogen peroxide) 2011, mc; *l*-**C₃H⁺** (propynylidyne cation) 2012, mc; **HMgNC** (hydromagnesium isocyanide) 2013, cs; **MgCCH** (magnesium monoacetylide) 2014, cs; **NCCP** (cyanophospha-ethyne) 2014, cs; **HCCO** (ketenyl) 2015, mc; **CNCN** (isocyanogen) 2018, mc; **HONO** (nitrous acid) 2019, sfr; **HCCS** (thioketenyl radical) 2021, sfr; **HNCN** (cyanomidyl radical) 2021, mc;

Molecules containing 5 atoms

HC₃N (cyanoacetylene) 1971, sfr; **HCOOH** (formic acid) 1971, sfr; **CH₂NH** (methanimine) 1973, sfr; **NH₂CN** (cyanamide) 1975, sfr; **H₂CCO** (ketene) 1977, sfr; **C₄H** (butadiynyl) 1978, cs; **SiH₄** (silane) 1984, cs; *c*-**C₃H₂** (cyclopropenylidene) 1987, sfr, mc; **CH₂CN** (cyanomethyl) 1988, sfr, mc; **C₅** (pentacarbon) 1989, cs; **SiC₄** (silicon tetracarbide) 1989, cs; **H₂CCC** (propadienylidene) 1991, mc; **CH₄** (methane) 1991, mc; **HCCNC** (isocyanoacetylene) 1992, mc; **HNCCC** 1992, mc; **H₂COH⁺** (protonated formaldehyde) 1996, sfr; **C₄H⁻** (butadiynyl anion) 2007, cs; **CNCHO** (cyanoformaldehyde) 2008, sfr; **HNCNH** (carbodiimide) 2012, sfr; **CH₃O** (methoxy) 2012, mc; **NH₃D⁺** (deuterated ammonium cation) 2013, sfr; **H₂NCO⁺** (protonated isocyanic acid) 2013, sfr; **NCCNH⁺** (protonated cyanogen) 2015, mc; **CH₃Cl** (chloromethane) 2017, cs; **MgC₃N** (magnesium moncyanoacetylide) 2019, cse; **NH₂OH** (hydroxylamine) 2020, mc; **HC₃O⁺** (protonated tricarbon monoxide) 2020, sfr; **HC₃S⁺** (protonated tricarbon monosulfide) 2021, sfr; **H₂CCS** (thioketene) 2021, sfr; **C₄S** (tetracarbonsulfide) 2021, sfr; *t*-**HC(O)SH** (*trans*-thioformic acid) 2021, sfr; **HCSCN** (cyanothioformaldehyde) 2021, sfr;

Molecules containing 6 atoms

CH₃OH (methanol) 1970, sfr; **CH₃CN** (methyl cyanide) 1971, sfr; **NH₂CHO** (formamide) 1971, sfr; **CH₃SH** (methyl mercaptan) 1979, sfr; **C₂H₄** (ethylene) 1981, cse; **C₅H** (pentynylidyne) 1986, cs; **CH₃NC** (methyl isocyanide) 1988, sfr; **HC₂CHO** (propynal) 1988, mc; **H₂CCCC** (butatrienylidene) 1991, cse; **HC₃NH⁺** (protonated cyanoacetylene)

1994, mc; C_5N (cyanobutadiynyl) 1998, mc; HC_4H (diacetylene) 2001, cs; HC_4N (no name for this radical) 2004, cs; c-H_2C_3O (cyclopropenone) 2006, sfr; CH_2CNH (ketenimine) 2006, sfr; C_5N^- (cyanobutadiynyl anion) 2008, cs; HNCHCN (E- and Z-cyanomethanimine) E-2013 and Z-2019, sfr; SiH_3CN (silyl-cyanide) 2014, cse; C_5S (pentacarbon monosulfide) 2014, cse; MgC_4H (magnesium monobuta diynide) 2019, cse; CH_3CO^+ (acetyl cation) 2021, sfr; CH_2CCH (propargyl radical) 2021, sfr; H_2CCCS (thiopropadienone) 2021, sfr; HCSCCH (propynethial) 2021, sfr;

Molecules containing 7 atoms

CH_3CHO (acetaldehyde) 1973, sfr; CH_3CCH (methyl acetylene) 1973, sfr; CH_3NH_2 (methylamine) 1974, sfr; CH_2CHCN (vinyl cyanide) 1975, sfr; HC_5N (cyanodiacetylene) 1976, sfr; C_6H (hexatriynyl) 1986, mc, c-C_2H_4O (ethylene oxide) 1997, sfr; CH_2CHOH (vinyl alcohol) 2001, sfr; C_6H^- (hexatriynyl anion) 2006, cs, mc; CH_3NCO (methyl isocyanate) 2015, sfr; HC_5O (butadiynylformyl) 2017, mc; HNCHCCH (propargylamine) 2020, mc; HC_4NC (isocyanodiacetylene) 2020, sfr; c-C_3HCCH (ethynyl cyclopropenylidene) 2021, sfr; H_2C_5 (pentatetraenylidene) 2021, sfr; MgC_5N (magnesium cyanodiacetylide) 2021, cse;

Molecules containing 8 atoms

$HCOOCH_3$ (methyl formate) 1975, sfr; CH_3C_3N (methylcyanoacetylene) 1984, mc; C_7H (heptatriynylidyne) 1997, mc; CH_3COOH (acetic acid) 1997, sfr; H_2C_6 (hexapenta-enylidene) 1997, mc; C_7H (heptatriynylidyne) 1997, cs; CH_2OHCHO (glycolaldehyde) 2000, sfr; HC_6H (triacetylene) 1997, cs; CH_2CHCHO (propenal) 2001, sfr; CH_2CCHCN (cyanoallene) 2006, mc; NH_2CH_2CN (aminoacetonitrile) 2008, sfr; $HCOOCH_3$ (methyl formate) c-1975 and t-2012, sfr; CH_3CHNH (ethanimine) 2013, sfr; CH_3SiH_3 (methyl silane) 2017, cs; $(NH_2)_2CO$ (urea) 2019, sfr; $HCCCH_2CN$ (propargyl cyanide), sfr; HC_5NH^+ (protonated cyanodiacetylene), 2020, sfr; CH_2CHCCH (vinyl acetylene) 2021, sfr; MgC_6H (magnesium triacetylide) 2021, cse;

Molecules containing 9 atoms

CH_3OCH_3 (dimethyl ether) 1974, sfr; CH_3CH_2OH (ethanol) 1975, sfr; CH_3CH_2CN (ethyl cyanide) 1977, sfr; HC_7N (cyanotriacetylene) 1978, mc; CH_3C_4H (methyl diacetylene) 1984, mc; C_8H (octatriynyl) 1996, cs; CH_3CONH_2 (acetamide) 2006, sfr; C_8H^- (octatriynyl anion) 2007, mc; CH_2CHCH_3 (propylene) 2007, mc; CH_3CH_2SH (ethyl mercaptan)

2014, sfr; **HC$_7$O** (hexadiynylformyl) 2017, mc; **CH$_3$NHCHO** (*N*-methyl-formamide) 2017, sfr; **HC≡CCH=CHC≡N** (*trans*-cyanovinyl acetylene) 2021, sfr; **H$_2$C=CHC$_3$N** (vinylcyanoacetylene) 2021, sfr; **H$_2$CCCHCCH** (allenyl acetylene) 2021, sfr;

Molecules containing 10 atoms

(CH$_3$)$_2$CO (acetone) 1987, sfr; **HO(CH$_2$)$_2$OH** (ethylene glycol) 2002, sfr; **CH$_3$CH$_2$CHO** (propanal) 2004, sfr; **CH$_3$C$_5$N** (methycyanodiacetylene) 2006, mc; **CH$_3$CHCH$_2$O** (propylene oxide) 2016, sfr; **CH$_3$OCH$_2$OH** (methoxymethanol) 2017, sfr; ***o*-C$_6$H$_4$** (*ortho*-benzyne) 2021, sfr; **H$_2$CCCHC$_3$N** (cyanoacetyleneallene) 2021, sfr;

Molecules containing 11 atoms

HC$_9$N (cyanotetraacetylene) 1978, mc; **CH$_3$C$_6$H** (methyltriacetylene) 2006, mc; **CH$_3$CH$_2$OCHO** (ethyl formate) 2009, sfr; **CH$_3$COOCH$_3$** (methyl acetate) 2013, sfr; **CH$_3$COCH$_2$OH** (hydroxyacetone) 2021, cse; **c-C$_5$H$_6$** (cyclopentadiene) 2021, sfr; **NH$_2$CH$_2$CH$_2$OH** (ethanolamine) 2021, mc;

Molecules containing 12 atoms

C$_6$H$_6$ (benzene) 2001, cs; **C$_3$H$_7$CN** (*n*- and i-propyl cyanide) *n*-2009, sfr, and i-2014, sfr; **C$_2$H$_5$OCH$_3$** (*trans* ethyl methyl ether) 2015, sfr; **c-C$_5$H$_5$CN** (1- and 2-cyanocyclopentadiene) 2021, sfr; **C$_2$H$_5$CONH$_2$** (propionamide) 2021, sfr;

Molecules containing 13 atoms

c-C$_6$H$_5$CN (benzonitrile) 2018, sfr; **HC$_{11}$N** (cyanopenta-acetylene) 2021, sfr;

Polycyclic aromatic hydrocarbon molecules (PAHs)

C$_{10}$H$_7$CN (1- and 2-cyanonaphthalene) 2021, sfr; **c-C$_9$H$_8$** (indene [fused benzene and cyclopentene rings]) 2021, sfr;

Cage molecules (fullerenes)

C$_{60}$ (buckminsterfullerene) 2010, cse; **C$_{60}$$^+$** (buckminsterfullerene cation) 2013, cse; and **C$_{70}$** (rugbyballene) 2010, cse.

References

1. B. A. McGuire, *Astrophys. J., Suppl.*, 2022, **259**, 30.
2. A. S. Eddington, *Proc. R. Soc. A*, 1926, **111**, 424–456.
3. T. Dunham, *Publ. Astron. Soc. Pac.*, 1937, **49**, 26.
4. P. Swings and L. Rosenfeld, *Astrophys. J.*, 1937, **86**, 483.
5. S. Weinreb, A. H. Barrett, M. L. Meeks and J. C. Henry, *Nature*, 1963, **200**, 829.
6. G. R. Carruthers, *Astrophys. J., Lett.*, 1970, **161**, L181.
7. R. W. Wilson, K. B. Jefferts and A. A. Penzias, *Astrophys. J., Lett.*, 1970, **161**, L43.
8. D. A. Williams and T. W. Hartquist, *The Cosmic-Chemical Bond*, Royal Society of Chemistry, 2013.
9. D. A. Williams and C. Cecchi-Pestellini, *Dust in Galaxies*, Royal Society of Chemistry, 2020.
10. www.astrochymist.org/astrochymist_ism.html.

2 Detecting Astronomical Molecules

"We report the results of the initial *Far Ultraviolet Spectroscopic Explorer* observations of molecular hydrogen in translucent clouds. These clouds have greater optical depth than any of the diffuse clouds previously observed for far-UV absorption and provide new insights into the physics and chemistry of such regions." (T. P. Snow *et al.*)[1]

2.1 Introduction

We've seen in Chapter 1 (Appendix) the large number and great variety of molecular species that have been detected in interstellar and circumstellar regions of the Milky Way and other galaxies. Interstellar species are mostly found in the cooler, denser, and more massive components of gas in the interstellar medium, while circumstellar species are found in the envelopes surrounding stars that are near the ends of their lives. Our main purpose in this book is to explore the chemistry that produces these molecules, but it seems appropriate before we start that task to describe the physics that enables these detections to be made.

It's rare for an astrochemist to have the opportunity to examine a piece of pre-terrestrial matter in a laboratory. Of course, in space

Astrochemistry: Chemistry in Interstellar and Circumstellar Space
By David A. Williams and Cesare Cecchi-Pestellini
© David A. Williams and Cesare Cecchi-Pestellini 2023
Published by the Royal Society of Chemistry, www.rsc.org

missions to solar system objects like the Moon, asteroids and comets, material from these objects can be collected and examined either remotely on board the spacecraft or returned to Earth and examined in terrestrial laboratories. Studies of that kind reveal much information about the formation of the Sun and the solar system. Earth itself acts as a collector of non-terrestrial matter; it is continually bombarded by interplanetary particles which can be found (particularly in snow-covered Antarctica), collected, classified, and analyzed. These particles reveal an interesting history that confirms signatures of an interstellar and circumstellar origin for many of these particles. In addition, Earth itself should be regarded not as a body distinct from the Universe but rather as a body formed from interstellar material that accumulated during the period when the Sun itself was forming, and has since undergone significant modification. Unravelling those modifying processes gives information about the pre-planetary phase of a solar system composed of material from interstellar clouds (as we'll see in Chapter 9).

The information on which the Appendix to Chapter 1 is based comes not from collected interplanetary particles but from *spectroscopy*, *i.e.*, the interaction of electromagnetic radiation and matter that leaves a specific characteristic signature of the matter on the radiation. The signature may be observed either in absorption or emission, depending on the local physical conditions; it may be at a particular wavelength (forming a line) or over a range of wavelengths (forming a continuum), and it may be in any part of the electromagnetic spectrum, from short wavelengths (X-rays and γ-rays) to long wavelengths (radio). The essential point is that the absorption or emission is a true signature of the species generating the interaction. The species may be an atom, an atomic ion, a molecule or a molecular ion, or matter in the solid state. In the following sections we discuss basic ideas about spectral line profiles (Section 2.2.1), spectra arising from molecular rotation transitions (especially important for almost all of the molecular species in the Appendix to Chapter 1) in Section 2.2.2, transitions involving molecular vibrations in Section 2.2.3, and we also make a few remarks about electronic spectra of molecules in Section 2.2.4, although electronic transitions are responsible for spectra that enabled the detection of only a very few interstellar molecules. Since many readers will be familiar with much of this material, it is summarized here only rather briefly. In Section 2.3 we introduce the concept of critical number density that determines which of the many possible rotational lines are likely to be prominent, and in Section 2.4

we discuss basic ideas of transfer of radiation through the interstellar medium. These ideas are dealt with in more detail in ref. 2.

2.2 Astronomical Spectroscopy

2.2.1 Spectral Lines and Line Profiles

A spectral line is formed when a species makes a transition between two energy states of a particular species. If energy is absorbed from a continuum radiation field, the transition is from the lower state to the upper (excited) state and an absorption line is formed in the continuum. Relaxation from the excited state to a lower state forms a line in emission, observed on top of any continuum.

A line is not infinitely narrow, but has a width. To be specific, let's consider an emission line. There are several mechanisms that contribute to the width of an emission line. *Natural line broadening* arises from Heisenberg's Uncertainty Principle: the amount of time that a species is in the upper state, Δt, and the uncertainty in the energy, ΔE, that will be emitted are related:

$$\Delta E \Delta t \geq \hbar \qquad (2.1)$$

where $\hbar$ is the Dirac constant (or $h/2\pi$, where h is the Planck constant), and $\Delta E = h\Delta v$ gives the uncertainty in frequency v. Thus, the uncertainty in the time of emission creates a range of frequencies, generating a width in the emission of a shape called a *Lorentzian line profile*, which in terms of v gives a shape to the intensity I of the form

$$I(v) \sim 1/[(v - v_0)^2 + (\gamma/4\pi)^2] \qquad (2.2)$$

where the line shape peaks at frequency v_0, the width is measured by γ which is identified with the Einstein A-coefficient for the transition (so $1/A$ is the occupation time for the upper level).

For strong transitions in the optical or ultraviolet part of the spectrum, A may be as large as $\sim 10^8$ s^{-1} so $\Delta v/v \sim 10^{-7}$, *i.e.*, the Lorentzian line width is very small. It is negligible for transitions in other parts of the spectrum, for which A coefficients are much smaller.

Of course, the source does not consist of a single atom or molecule, but is a gas of atoms and molecules in motion with a random range of velocities. The radiation emitted at frequency v_0 is Doppler-shifted to frequency v because of the component of the velocity in the direction of travel of the photon, say v_z, according to the equation

$$(v - v_0)/v_0 = v_z/c \qquad (2.3)$$

In most cases, the velocity range of atoms or molecules in the gas is Maxwellian so the number of emitters with velocities in the range v_z to $(v_z + dv_z)$ is proportional to $\exp(-Mv_z^2/2kT)dv_z$, where M is the mass of the emitting atom or molecule. This gives the emission a so-called *Gaussian line profile*, expressed in frequency of

$$I(v) \sim \exp[-(v - v_0)^2/2\delta^2] \qquad (2.4)$$

where

$$\delta^2 = v_0^2 kT/Mc^2 \qquad (2.5)$$

This Gaussian (or Doppler) profile is more strongly peaked than the Lorentzian. The Gaussian obviously represents the gas at temperature T. Turbulent effects, as distinct from thermal temperatures, will also contribute to the profile *via* an effective temperature higher than the actual thermal temperature. If the gas is in motion relative to Earth, then all the frequencies are shifted in the way described in eqn (2.3). If there are several clumps, at different temperatures and with different velocities along the line of sight, all of these will contribute separately to the line profiles. Thus, spectroscopy has the potential to reveal density and temperature structure within a cloud, and the possible presence of turbulence.

Collisional Broadening occurs when collisions interrupt and shorten the time available for emission. As in natural line broadening, the broadening effect is described by a Lorentzian profile. However, this effect is insignificant in the interstellar medium where the densities are so low that collisions are very infrequent. It can be significant in much denser regions, such as planetary atmospheres.

There are also a number of other types of line profile in which several of the above broadening mechanisms are convolved. For example, the Voigt line profile is a convolution of Doppler broadening with collisional broadening. These profiles have been developed because they have applications in different physical regimes.

2.2.2 Molecular Spectra of Interstellar Molecules: Rotational Spectra at Millimetre Wavelengths

Linear Molecules

The simplest model of a rotating *diatomic* molecule is a *rigid rotator*, considered to be two point-like masses, m_1 and m_2, fixed to a rigid mass-less rod of length r. Classically, the energy of the rotating rigid body is $E = \frac{1}{2}I\omega^2$, where I is the moment of inertia and ω is the angular momentum (ω is $2\pi\nu_{rot}$, where ν_{rot} is the rotational frequency of the rotator). The moment of inertia of this system is simply

$$I = m_1 r_1^2 + m_2 r_2^2 \qquad (2.6)$$

where $r_1 + r_2 = r$ and the centre of mass of the rotator is in the rod at a distance r_1 from m_1 and r_2 from m_2, so that

$$r_1/r = m_2/(m_1 + m_2); \quad r_2/r = m_1/(m_1 + m_2) \qquad (2.7)$$

Classically, the rigid rotator can have any energy E, but the solution of the Schrödinger equation for the rigid rotator shows that the energy of a *quantum mechanical rigid rotator* such as a diatomic molecule is restricted to certain values E_J where

$$E_J = BJ(J + 1) \qquad (2.8)$$

where J is the rotational quantum number; it takes integer values 0, 1, 2, 3, ... Here,

$$B = (\hbar^2/2I) \qquad (2.9)$$

is called the rotational constant; it has units of energy. The energy levels permitted for a quantum mechanical rigid rotator are, therefore,

0, 2B, 6B, 12B, ... and the transitions between these permitted levels correspond to energies (and, therefore, frequencies) 2B, 4B, 6B, ... corresponding to transitions in quantum number J of 1-0, 2-1, 3-2, *etc.* So the spectrum of a quantum mechanical rigid rotator is a set of lines whose frequencies increase by steps of 2B.

These results apply not only to *diatomic* molecules but also to *linear polyatomic* molecules such as cyanodiacetylene HC_5N (which has the linear structure $H–C\equiv C–C\equiv C–C\equiv N$).

Evidently, since B has units of energy, and for the molecule CO (carbon monoxide), B is 3.83×10^{-23} Joules, the 1-0 transition for this molecule has a wavelength of 2.6 mm and the 2-1 transition has a wavelength of 1.3 mm, *etc.* For less massive molecules, the moment of inertia is smaller so B is larger, along with larger energies and frequencies, so wavelengths are shorter, possibly less than one mm. The wavelength of the 1-0 rotational transition in CH is about 0.3 mm. Similarly, more massive molecules have larger moments of inertia so their rotational constants are smaller, and their rotational energy levels are closely spaced and transitions have lower energy and longer wavelengths. For example, the linear molecule cyanodiacetylene (HC_5N) has a 4-3 rotational transition with wavelength 29 mm.

The rigid rotator model cannot be an exact representation of a diatomic molecule, because the length of the rod joining the atoms of a diatomic molecule must be able to change when the molecule is capable of vibrating. So the length of the rod changes as the molecule rotates faster and faster. A model known as the *nonrigid rotator* attempts to allow for this by connecting the atoms of the diatomic molecule by a mass-less spring which becomes extended at higher rotational frequencies. This centrifugal distortion results in an extra term, D, in the energy formula

$$E_J = BJ(J + 1) - D[J(J + 1)]^2 \tag{2.10}$$

However, D is found to be generally very much less than B, so this centrifugal distortion is usually unimportant.

Symmetric Top Molecules

While the Appendix to Chapter 1 includes some diatomic and linear molecules, most of the species listed do not fall into these categories; they are structurally more complex. We consider next molecules called *symmetric top molecules*. In these molecules, two of the three principal moments of inertia (here called I_A, I_B, and I_C)

are equal (say, $I_B = I_C$). We can discuss their rotational spectra as an extension of the spectra of linear molecules.

For symmetric top molecules, the energy of rotation includes rotation about the axis of symmetry (for which the moment of inertia is, say, I_A) and rotation perpendicular to that axis (for which there are two equal moments of inertia, say, I_B and I_C). We require two quantum numbers, J for the total angular momentum and K for the projection of the total angular momentum on to the symmetry axis. By analogy with the diatomic case, we can write

$$E(J,K) = BJ(J + 1) + (A - B)K^2 \qquad (2.11)$$

where, similarly to eqn (2.9),

$$B = \hbar^2/2I_B \text{ and } A = \hbar^2/2I_A \qquad (2.12)$$

If $A > B$, the molecule is called *prolate* (tending towards a needle-shape), and if $A < B$, the molecule is called *oblate* (tending towards a disc-shape). Since the projection of the total angular momentum cannot exceed the total, K cannot exceed J so the values that K may take are

$$K = -J, -J + 1, -J + 2, -J + 3 \dots J$$

with selection rules

$$\Delta J = 0, \pm 1, \text{ and } \Delta K = 0.$$

Asymmetric top molecules have rotational motion that is even more complicated than for symmetric tops, and lies between the extremes of oblate and prolate tops. No simple formulae can be given. Nearly all the molecules in the Appendix to Chapter 1 are asymmetric tops.

Let's suppose that the asymmetric top has principal moments of inertia I_A, I_B, and I_C, and that A, B, and C are defined similarly as in eqn (2.12). One way to proceed is to calculate the energy levels for an oblate symmetric top ($I_B = I_A$). Letting I_B increase gradually from $I_B = I_A$ to $I_B = I_C$ (a prolate symmetric top), we find a continuous change in energy levels from the levels of a given value of J on one side of this change to the same value of J on the other side. For example, the energy levels for an oblate symmetric top with $J = 2$, $K = (0, 1, 2)$ become the energy levels for a prolate symmetric top with $J = 2$, $K = (2, 1, 0)$, passing through energy states for the asymmetric top labelled as 2_{+2}, 2_{+1}, 2_0, 2_{-1}, and 2_{-2}. The notation used for these states is either as shown, or the limiting K values are shown as a subscript to the appropriate J value: 2_{20}, 2_{10}, 2_{00}, 2_{02}, 2_{01}, and 2_{02}. Accurate values of the corresponding wavelengths or frequencies must be obtained from laboratory measurements of the molecular spectra.

Observed Interstellar Frequencies for Rotational Transitions

Most of the species listed in the Appendix to Chapter 1 are detected by emission in rotational transitions. The detected frequencies are usually in the range of a hundred to a thousand GHz, corresponding to wavelengths of about 3 to 0.3 mm. For example, as we have seen, the near-ubiquitous CO (1-0) rotational transition is at 115.27 GHz (a wavelength of 2.6 mm). The CO (6-5) rotational transition is at 691.47 GHz (0.43 mm). Cyanoacetylene (HC_3N) has its 12-11 rotational transition at 109.17 GHz (2.75 mm) while its 19-18 rotational transition is at 172.85 GHz (1.73 mm).

However, although the detection of interstellar molecules by their rotational spectra has been phenomenally successful, there are some limitations to the use of rotational spectra in tracing or detecting interstellar molecules. The strength of a rotational transition is proportional to the square of the permanent dipole moment in the molecule, so if the dipole moment is very weak or zero, then the molecule does not show a useful rotational spectrum. Thus, molecules such as H_2, N_2, and CH_4 (all of which have zero dipole moments) do not show a rotational spectrum in the interstellar medium. Also, for larger species with non-zero dipole moments, the number of rotational levels accessible at low temperatures becomes very large (because, as we have seen, the rotational constants are small for large molecules). Therefore, the number of large molecules emitting or absorbing radiation at a particular frequency becomes very small. Consequently, the emission in any particular transition is very weak and may be undetectable.

2.2.3 Molecular Spectra of Interstellar Molecules: Vibrational Spectra at Infrared Wavelengths

Diatomic and Polyatomic Molecules

The vibration of a diatomic molecule with atomic masses m_1 and m_2 can be represented by a *harmonic oscillator* in which a point mass m, free to move along the x-axis, is acted on by a force F directed always towards the equilibrium position, $F = -kx = m d^2x/dt^2$. This force is generated by a potential energy $V = \frac{1}{2}kx^2$ (a parabola). Classically, this oscillator may have any energy, and has a natural frequency of oscillation

$$v_{osc} = (1/2\pi)(k/m)^{1/2} \qquad (2.13)$$

The equations of motion of the two masses, m_1 and m_2, of the diatomic molecule, at distances r_1 and r_2 from the centre of mass, can be combined in the equation

$$\mu d^2(r - r_e)/dt^2 = k(r - r_e) \qquad (2.14)$$

where μ is the reduced mass, $\mu = m_1 m_2/(m_1 + m_2)$, and r and r_e are the instantaneous and equilibrium separations between the atoms, respectively, and $r = r_1 + r_2$. The eqn (2.14) is identical with that of the harmonic oscillator, so the vibration of a diatomic molecule can be considered as a harmonic oscillator with natural frequency as in (2.12) but with m replaced by μ.

Quantum mechanically, however, the oscillator is not confined to the single classical natural frequency, but a range of energies is permitted:

$$E(v) = h v_{osc}(v + \tfrac{1}{2}) \qquad (2.15)$$

where v is the vibrational quantum number ($v = 0, 1, 2, ...$) and h is the Planck constant, and transitions are subject to the selection rule that in a transition v must change (up or down) by 1. The equation shows that the state of lowest energy, corresponding to $v = 0$, is not zero but $\tfrac{1}{2}h v_{osc}$. This energy is called the *zero point energy*. Eqn (2.15) shows that the energy levels for the harmonic oscillator are spaced equally by an amount $h v_{osc}$. The numerical value of this term can be determined from spectra.

In reality, a harmonic oscillator is not a perfect model for vibration in diatomic molecules, especially for higher vibrational excitation where the potential energy departs very significantly from a parabola, see Figure 2.1.

A simple correction is to include additional terms in eqn (2.13), $(v + \tfrac{1}{2})^2$ and $(v + \tfrac{1}{2})^3$. This model is called the *anharmonic oscillator*. The contribution of these terms must be determined experimentally from spectra.

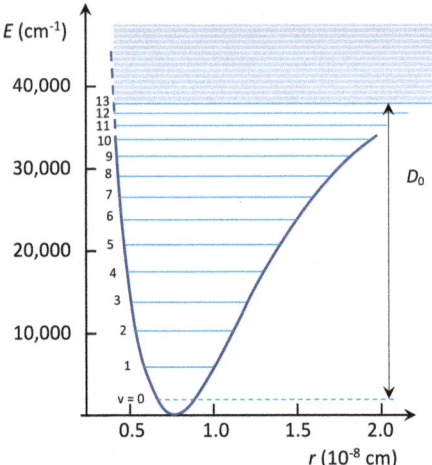

Figure 2.1 The potential energy curve for the ground state of molecular hydrogen, showing the departure from the parabolic form. The figure shows also the vibrational energy levels for this molecule. Note how the levels become closer together at higher excitation. Above the highest vibrational level (v = 14) is a continuous spectrum (shown shaded). In this continuum the two H-atoms are unbound. The energy D_0 from v = 0 to the beginning of the continuum is the dissociation energy, shown in units of inverse cm, where $10\,000$ cm^{-1} = 1.24 eV.

Polyatomic molecules have many vibrational modes but these can be represented by a collection of harmonic oscillators, each with its own characteristic motion and oscillation frequency. We shall not consider vibrations within these molecules separately.

Wavelengths of vibrational transitions are typically in the range of a few microns and above, *i.e.*, in the infrared region of the spectrum. For example, wavelengths of the 1-0 vibrational transitions in HD, CH, CO, and CN are approximately 2.62, 3.49, 4.61, and 4.84 μm, respectively. Vibrational lines in emission arise in regions that are much warmer than the typical temperatures of dark interstellar clouds, to temperatures of hundreds to thousands of Kelvin, so thermal collisions are sufficiently energetic to excite vibrational transitions, and the populated levels relax and the line appears in emission. If a suitable background source of infrared continuum exists, then infrared lines may be seen in absorption in the interstellar medium. Vibrational lines in absorption have been useful in detecting symmetric molecules that have no (or weak) dipole moments and are invisible in rotational spectra.

Fairly simple interstellar molecules are also found in solid ices on interstellar grains (as we shall discuss in Chapter 8). The rotational motions of these embedded molecules are suppressed in this environment, but the molecules may be detected by pure vibrational spectra with no rotational structure. Some modification of the frequencies and line profiles may occur in the pure vibrational spectra. The laboratory spectroscopy of mixed ices has become a necessary expertise to identify molecules embedded in interstellar ices of a variety of compositions.

2.2.4 Molecular Spectra of Interstellar Molecules: Electronic Spectra in the Optical and Ultraviolet

If they can be observed, electronic transitions in interstellar molecules appear in absorption rather than emission. The conditions required to raise molecules into excited electronic states (typically about 10 eV above the ground electronic state) from which *emission* will occur are sufficiently extreme (for example, those arising in hot stellar atmospheres) that molecules under those conditions will be rapidly destroyed. Indeed, as we have seen in Section 1.1.1, the first three detections of interstellar molecular species in the Milky Way were made in optical *absorption* by interstellar molecular species. A background bright star is a source of optical and ultraviolet light, and the absorbing species are in an interstellar cloud in the line of sight from Earth to that star.

While the number of interstellar molecular detections *via* electronic absorptions is small, one of these detections is particularly important for astrochemistry. Interstellar molecular hydrogen was first detected in its ultraviolet absorption spectrum against a hot star as a radiation source. The detection was made by a rocket-borne spectrometer (see Chapter 1, ref. 6). The spectrometer obtained the ultraviolet absorption spectrum of molecular hydrogen in the Lyman bands $(B^1\Sigma_u^+ \leftarrow X^1\Sigma_g^+)$ at wavelengths near 100 nm, from the ground electronic and vibrational level to vibrational levels of the upper electronic state, B, from $v = 0\text{--}7$. The background star was ξ Persei, a very hot (effective temperature 35 000 K) and bright (0.26 million solar luminosities) star. The observation inferred that about half of the interstellar hydrogen on the line of sight towards ξ Persei was molecular. The Lyman $(B^1\Sigma_u^+ \leftarrow X^1\Sigma_g^+)$ and Werner $(C^1\Pi_u \leftarrow X^1\Sigma_g^+)$ absorption bands (here, B and C are the first two stable excited electronic states) were subsequently observed towards the star δ Scorpii from rocket-borne observations in 1973.[3] Detailed studies of interstellar molecular hydrogen were subsequently made using Earth-orbiting satellites, especially the Far Ultraviolet Spectroscopic Explorer (FUSE) mission which operated during 1999–2007. Figure 2.2 shows a small portion of the absorption spectrum[1] taken by FUSE of molecular hydrogen at wavelengths near 100 nm in the far ultraviolet, this molecular hydrogen being located along the line of sight towards a very bright star named HD73882. We shall discuss these spectra and their relation to the photodissociation of molecular hydrogen in the next chapter.

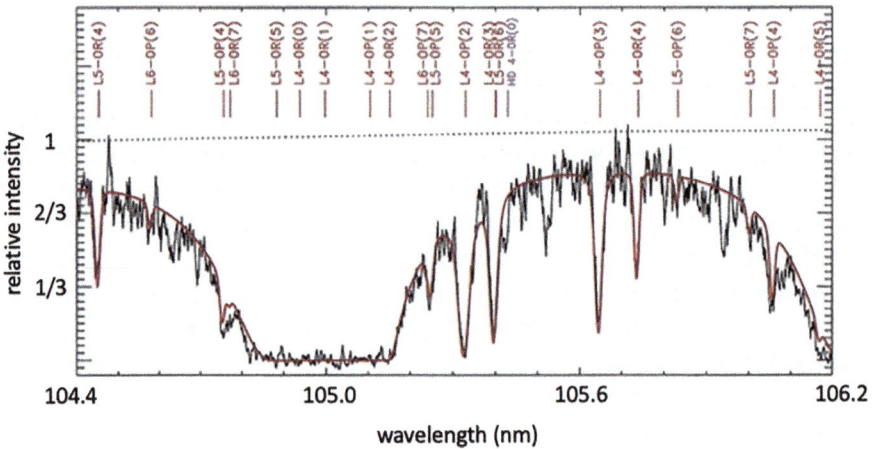

Figure 2.2 A portion of the absorption spectrum of H_2 at wavelengths between 104 and 106 nm. H_2 lies along the line of sight towards the bright star HD73882, the background source. The spectrum shows vibrational and rotational substructure in the electronic $B^1\Sigma_u^+ \leftarrow X^1\Sigma_g^+$ transition. In the figure shown, absorption occurs from individual rotational levels (J = 1–7) of the ground vibrational level (v = 0) of the electronic ground state X into various rotational levels of the several vibrational levels (v = 4, 5, and 6) of the upper electronic state B. The notation in the figure for rotational transitions is illustrated as follows: R(4) refers to a rotational transition J = 4 to J = 5, and P(6) refers to a rotational transition from J = 6 to J = 5. More generally, transitions in which ΔJ = −2, −1, 0, +1, and +2 are referred to by the symbols O, P, Q, R, and S, respectively. Thus, the line of shortest wavelength shown in the diagram (104.45 nm), designated L5-OR(4), corresponds to the electronic Lyman transition (L) in which v changes from 0 to 5 and J from 4 to 5. Reproduced with permission from ref. 1, https://iopscience.iop.org/article/10.1086/312791, under the terms of the CC BY 4.0 licence.

2.3   Critical Number Density

We've seen that there are many molecular species existing in inter-stellar space and in circumstellar regions. We also know that each of these species may, if the physical conditions are appropriate, emit or absorb line radiation. Can we observe all of these lines, or are some of them more prominent – depending on the local physical conditions?

For simplicity, let's consider a linear molecule X with a rotational spectrum as described in Section 2.2.2 and let's consider the two lowest energy states corresponding to $J = 0$ and $J = 1$. In an interstellar molecular cloud there are three important processes connecting these states. First, collisions with the most abundant species, molecular hydrogen, may excite the molecule into the excited state:

$$X(J = 0) + H_2 \rightarrow X(J = 1) + H_2 \qquad (2.16)$$

where the energy of rotation in $X(J = 1)$ comes from the kinetic energy of the H_2 molecule. Secondly, the excited molecule X may then radiate spontaneously at a rate equal to the Einstein A-coefficient for the (1-0) transition in molecule X:

$$X(J = 1) \rightarrow X(J = 0) + h\nu \qquad (2.17)$$

where $h\nu$ is the emitted photon whose energy came originally from the thermal energy of the H_2 molecule. If this photon escapes from the gas, this is a cooling process. Alternatively, if the number density of H_2 is sufficiently great, collisions of H_2 with $X(J = 1)$ may remove the energy from X before the molecule has had a chance to radiate:

$$X(J = 1) + H_2 \rightarrow X(J = 0) + H_2 + \Delta E \qquad (2.18)$$

where ΔE is the rotational energy now restored to kinetic energies of the partner molecules. Whether or not the 1-0 rotational line is emitted depends on the relative rates of the reactions (2.16), (2.17), and (2.18). If the H_2 number densities are sufficiently high, then nearly all excitations (2.16) are followed rapidly by collisional de-excitations (2.18) so very few, if any, photons are emitted. On the other hand, if the H_2 number density is sufficiently low then few excitations (2.16) are made so few rotational photons are emitted (2.17).

However, if the rates of (2.16) and (2.17) are comparable, then the maximum emission occurs. This statement can be written as an approximate equality

$$n(H_2)\langle Q_{10}v(H_2)\rangle \sim A_{10} \tag{2.19}$$

where the left hand side is the rate of (2.16), in which the product of the cross section Q_{10} and velocity $v(H_2)$ is integrated over all velocities, and on the right hand side is the rate of (2.17), *i.e.*, the Einstein *A*-coefficient for spontaneous emission of molecule $X(J = 1)$ in the 1-0 transition. The number density for which the approximate equality holds in (2.19) is called the *critical number density* for emission by molecule X in the $1 \to 0$ transition. This critical density is calculated from assumed equality in (2.19), for an adopted temperature. The molecular data necessary for calculations like these are maintained at the website home.strw.leidenuniv.nl/~moldata/ (see also ref. 4). For example, for carbon monoxide at 10 K, the critical number density for the 1-0 rotational transition is 1.8×10^3 cm^{-3} so this transition is useful for probing molecular clouds with density and temperature in those ranges. We can generalize eqn (2.19) so that it applies to any transition $m \to n$ rather than $1 \to 0$:

$$n(H_2)\langle Q_{mn}v(H_2)\rangle \sim A_{mn} \tag{2.20}$$

For example, for CO at a temperature of 100 K, the critical number density for the 6-5 rotational transition is 2.5×10^5 cm^{-3}. Evidently, quite different regions of interstellar clouds are probed by different molecular transitions. No single transition is suitable to use for probing all types of region. Note that collisions with species other than H_2 may be important. It is known that collisions with electrons can reduce significantly the critical densities computed according to eqn (2.20). A few critical densities computed according to (2.20) are given in Table 2.1.

2.4   Radiative Transfer

In this chapter we have described (in Section 2.2) how a spectrum of lines may arise from molecules in the interstellar gas, and we have shown (in Section 2.3) how to select those lines that are likely to be

Table 2.1 Critical densities of a few selected species at specific temperatures.

Molecule	Formula	Transition	Frequency (GHz)	n_{crit} (cm^{-3})	T (K)
Carbon monoxide	CO	1-0	115.27	1.8×10^3	10
Carbon monoxide	CO	6-5	691.47	2.5×10^5	100
Formyl cation	HCO$^+$	2-1	178.38	1.1×10^6	10
Formyl cation	HCO$^+$	7-6	624.21	4.9×10^7	100
Cyanoacetylene	HC$_3$N	12-11	109.17	7.1×10^5	20
Cyanoacetylene	HC$_3$N	19-18	172.85	2.9×10^6	80

most useful in probing the particular physical conditions of that gas. Of course, the intensity of the beam as it travels through the gas is affected by emission and absorption of radiation by molecules in the gas. That is the topic of this section.

Let's consider the case of an isolated molecular cloud at a temperature that is high enough so that collisions are able to excite a molecule to the upper state of a rotational transition, say, CO $J = 1$. Then as long as the H$_2$ number density is not too high so that the excited rotational state is collisionally de-excited (as we discussed in Section 2.3), the molecule emits radiation and although some of the radiation emitted by one molecule may be re-absorbed by another, the cloud is a source of emission in that line (say, CO $J = 1$–0, for example). The total energy emitted into all solid angles per unit volume per second is simply $h\nu_{10} n_1 A_{10}$, where A_{10} is the Einstein spontaneous transition probability between levels 1 and 0, and n_1 is the number density of molecules in excited level 1. In thermodynamic equilibrium, the population in level i, say, n_i, will be given by

$$n_i = n_0 \exp(-\varepsilon_i/kT) \tag{2.21}$$

where n_0 is the number density of molecules in level 0 and T is the kinetic temperature. However, departures from thermodynamic equilibrium can occur. Nevertheless, the intensity of the emitted radiation can in principle be used to determine the number density and temperature of the emitting molecules.

A more complex situation occurs when radiation from a source (such as an interstellar cloud) passes through another interstellar cloud on its passage to Earth. Do astronomers on Earth see an emission line or an absorption line? To examine this situation in more detail we need to consider the equation of radiative transfer; this can be written as follows:

$$dI(v)/ds = -\kappa(v)I(v) + j(v) \tag{2.22}$$

Here, $I(v)$ is the intensity, defined as the energy per second in the frequency range v, $v + dv$ crossing unit area in unit solid angle; $\kappa(v)$ is the absorption coefficient per unit length of path, s; and $j(v)$ is the emissivity defined so that $j(v)dV\,dv\,d\Omega\,dt$ is the energy emitted by volume element dV in the frequency width dv in time interval dt into solid angle $d\Omega$. Astronomers measure absorption at frequency v by optical depth τ_v, where $d\tau_v = \kappa(v)ds$. We can re-write the equation of radiative transfer with τ_v as the independent variable:

$$dI(v)/d\tau_v + I(v) = j(v)/\kappa(v) \tag{2.23}$$

This equation is linear and may be integrated to give

$$I(v) = I_0(v) + \int_0^{\tau_v} \frac{j(v)}{\kappa(v)} \exp\left[-(\tau_v - \tau_v')\right] d\tau_v' \tag{2.24}$$

where $I_0(v)$ is the initial intensity. In molecular clouds it is often the case that the emission is determined by the collisions that populate the upper level. If Kirchhoff's Law applies, then

$$j(v) = \kappa(v)B_v(T) \tag{2.25}$$

where $B_v(T)$ is the Planck function

$$B_v(T) = (2hv^3/c^2)[\exp(hv/kT) - 1]^{-1} \tag{2.26}$$

and the solution of the equation of radiation transfer may be written as

$$I(v) = I_0(v)e^{-\tau_v} + B_v(T)\left(1 - e^{-\tau_v}\right) \tag{2.27}$$

Radio astronomers use the concept of *brightness temperature*, T_b, defined as the temperature of the black-body which would give the same intensity $I(v)$ in the frequency range observed. At radio wavelengths, $hv/kT \ll 1$ and so $e^{hv/kT} - 1 \approx hv/kT$ so that

$$B_v(T) \approx 2v^2kT/c^2 \tag{2.28}$$

(which is, in fact, the Rayleigh–Jeans Law). Upon defining T_b^0 to be the brightness temperature of the source,

$$T_b = T_b^0 e^{-\tau_\nu} + T\left(1 - e^{-\tau_\nu}\right)$$

(2.29)

is the brightness temperature measured; T is the temperature defining the populations in the atomic levels involved. If $T_b^0 > T$, then an absorption line appears in the $T_b - \nu$ diagram. For, if τ_ν is small at the line centre, then

$$T_b \approx T_b^0\left(1 - \tau_\nu\right) + T\tau_\nu = T_b^0 - \left(T_b^0 - T\right)\tau_\nu$$

(2.30)

which shows that T_b decreases as ν approaches the line centre. If τ_ν is large at the line centre, then $e^{-\tau_\nu}$ may be neglected there and so $T_b \approx T$ but it is larger than this outside the line. The area under the line in the $T_b - \nu$ diagram is proportional to the number of absorbers in the column. Evidently, these equations determine whether a line in the radio regions of the spectrum is seen in emission or absorption. A general introduction to radiative transfer in astrophysics may be found in ref. 5.

The *emission* of radiation when τ_ν is small is merely the sum of the emissions per atom. Thus, as we noted above, the total energy emitted into all solid angles per unit volume per second is $h\nu_{10}n_1A_{10}$, where A_{10} is the Einstein spontaneous transition probability between levels 1 and 0, and n_1 is the number density in the excited level 1. The line shape will be that described earlier, and the total energy is the total inside the profile. The problem is in the calculation of n_1. This may be given by a thermodynamic description, in which the number in level i at energy ε_i is given by eqn (2.21), where T is the kinetic temperature. Sometimes, however, T is not the kinetic temperature, but a temperature characterizing the mechanisms controlling the level populations; in this case, it is called the *excitation temperature*. Many emission lines from interstellar molecules, in particular, are found to arise in situations which are far from thermodynamic equilibrium. In these cases, the molecules are radiatively or collisionally 'pumped' so the population n_1 in the excited level involved is considerably greater (for example, OH) or considerably less (for example, H_2CO) than the expected value.

How do we know that this is so? Let us consider the example of radio emission from interstellar OH, at a wavelength near 18 cm. If we interpret observations of OH to give the populations of the levels, and assume these populations have arisen as a result of a thermal process, then we may have a very surprising answer, for the implied temperature from eqn (2.21) can be enormous. For OH in some astronomical sources in which the line is in emission, the inferred temperature may exceed 10^9 K. This is obviously very much greater than can actually be the case, for the molecules could not survive at such temperatures. In addition, the line radiation is observed to be highly polarized and rapidly variable, both phenomena indicating maser action. To establish a maser in a three-level system between levels 1 and 0 requires the system to be 'pumped' to some third level, 2, and to cascade quickly into level 1, building a large (non-thermal) population in that level. The intensity of the maser in the 1-0 transition is therefore linked to the pumping mechanism, which—if it can be identified—will give information about the radiation intensity at the frequency of the pump, or about the total gas density if the pumping is by collision.

Spectroscopic studies clearly provide a precise method of identifying molecular species in interstellar and circumstellar space. More detailed studies based on the ideas indicated in this section may also give information on the distribution of the molecules and their physical environments. All these data feed into the development of models of interstellar and circumstellar chemistry to be described in the following chapters.

References

1. T. P. Snow, B. L. Rachford, J. Tumlinson, J. M. Shull, D. E. Welty, W. P. Blair, R. Ferlet, S. D. Friedman, C. Gry, E. B. Jenkins, A. Lecavelier, M. Lemoine, D. C. Morton, B. D. Savage, K. R. Sembach, A. Vidal-Madjar, D. G. York, B.-G. Andersson, P. D. Feldman and H. W. Moos, *Astrophys. J., Lett.*, 2000, **538**, L65.
2. B. T. Draine, *Physics of the Interstellar and Intergalactic Medium*, Princeton University Press, 2011.
3. A. M. Smith, *Astrophys. J., Lett.*, 1973, **179**, L11.
4. F. L. Schoier, F. F. S. van der Tak, E. F. van Dishoeck and J. H. Black, *Astron. Astrophys.*, 2005, **432**, 369.
5. G. B. Rybicki and A. P. Lightman, *Radiative Processes in Astrophysics*, WILEY-VCH Verlag GmbH & Co. KGaA, 2004.

3 Gas Phase Reactions in Interstellar and Circumstellar Media

"Interstellar chemistry is dominated by kinetics rather than thermodynamics. This reflects the highly non-equilibrium nature of the interstellar medium as low gas phase temperatures and densities combine with high fluxes of UV photons or energetic ions. In such a system, reactions and their rates are key to understanding the chemical composition of the medium." (A. G. G. M. Tielens)[1]

3.1 Introduction

In Section 1.1.3 we described very briefly the typical physical conditions in interstellar and circumstellar regions in which molecules are found. The gas number densities in these regions are generally very low or even extremely low compared to many other types of region in which more familiar chemistry occurs. The typical temperatures are also generally very low indeed in interstellar molecular regions (although high in circumstellar regions), and so typical pressures in an interstellar molecular cloud may be as low as $\sim10^{-17}$ of the standard pressure in Earth's atmosphere. Obviously, we need to consider very carefully how chemistry is able to proceed in such tenuous

Astrochemistry: Chemistry in Interstellar and Circumstellar Space
By David A. Williams and Cesare Cecchi-Pestellini
© David A. Williams and Cesare Cecchi-Pestellini 2023
Published by the Royal Society of Chemistry, www.rsc.org

regions. While temperatures in circumstellar regions are generally much higher than in interstellar clouds, the pressures in circumstellar regions are probably less than one millionth of Earth's standard atmospheric pressure. As pointed out in ref. 1 (and discussed also in Chapter 5), chemistry in gas at very low density and very low temperature subjected to ultraviolet and energetic particle radiation is determined by kinetics rather than thermodynamics. In this chapter, therefore, we discuss the main types of reaction that may occur in interstellar and circumstellar media.

These reactions may be important in the chemical networks that generate the detected interstellar and circumstellar species as listed in the Appendix to Chapter 1. We shall discuss in Chapter 4 the actual chemical networks that are believed to be important in producing the observed molecules in the various types of interstellar and circumstellar media. Our aim in this chapter is to describe briefly the physical principles enabling these processes to occur. A useful survey of these processes – and many other aspects of astrochemistry – can be found in ref. 1.

3.2 Gas Phase Chemistry in Interstellar Clouds

3.2.1 Radiative Association

The simplest possible mechanism of molecule formation is that two atoms, A and B, simply collide and bond together to form a diatomic molecule. However, the collisional complex AB* formed in the collision still contains enough energy for dissociation to occur (*i.e.*, for the atoms simply to "bounce" apart), so the complex needs to lose energy while the two atoms are in contact if the complex is to be stabilized.

There are two ways in which this energy loss may happen: the complex may radiate energy away, represented here by a photon of energy $h\nu$,

$$A + B \rightarrow AB^* \rightarrow AB + h\nu$$

in the process known as *radiative association*, or the complex may be stabilized by collision with a third body, M, another atom or molecule; if the third body takes energy away from the colliding pair (M*), then the atoms cannot separate, and the complex will be stabilized.

$$A + B + M \rightarrow AB^* + M \rightarrow AB + M^*$$

This process is called a *three-body (or termolecular) reaction*. We'll postpone the discussion of three-body reactions until Section 3.2.11 below.

The difficulty in making molecules *via* radiative association is that the time that A and B are in contact is very short, perhaps comparable to the time of one vibrational period in the complex. This period may be $\sim 10^{-13}$ s, or even shorter. On the other hand, the probability of radiating is at best (for strong transitions in the ultraviolet) $\sim 10^8$ s^{-1} and is likely to be very much slower. Thus, the fraction of collisions during which radiation occurs is very small ($\sim 10^{-13}$ s $\times \sim 10^8$ s$^{-1} \sim 10^{-5}$) and the reaction rate coefficient for forming molecules by radiative association of two atoms is not likely to be an efficient route.

If one of the partners is ionized, then the interaction becomes stronger because the charge induces a dipole in the neutral partner. The attraction between the charge and the induced dipole creates a longer range attraction than in the neutral case, so the rate coefficient is larger in the ionized case. For example, in the radiative association of C^+ and H to form CH^+,

$$C^+ + H \rightarrow CH^+ + h\nu,$$

the rate coefficient is $\sim 1.7 \times 10^{-17}$ cm^3 s^{-1} for temperatures in the range 10–300 K.

The radiative association of H atoms and protons

$$H + H^+ \rightarrow H_2^+ + h\nu$$

is a viable mechanism with a rate coefficient of $\sim 10^{-18}$ cm^3 s^{-1} at temperatures of a few hundred kelvin. It could be an initial step towards chemical complexity, but in fact it is too slow to have a significant effect on chemistry in the Milky Way.

However, the significance of radiative association may change considerably if one or both of the reacting partners are not single atoms or ions but *radicals* R_1 and R_2, at least one of which comprises several atoms. Some reactions of the type

$$R_1 + R_2 \rightarrow R_1R_2^* \rightarrow R_1R_2 + h\nu$$

are known to proceed without activation energy and at about the collisional rate if no bonds are broken. When formed, the complex $R_1R_2^*$ obviously still has sufficient energy to dissociate, but if the rate of dissociation is slow enough to allow the complex to radiate then the stable molecule R_1R_2 may form. The complex may be considered as a system of loosely-coupled oscillators between which energy may flow. The bond energy released in the formation of $R_1R_2^*$ is first shared among these oscillators. The probability that this energy will be returned to the same bond (causing the complex to dissociate) is low, lengthening the lifetime of the complex. The lifetime can if necessary be estimated using RRKM (Rice–Ramsperger–Kassel–Marcus) theory. This theory shows that the lifetime of the complex increases very rapidly with size of the system, so complexes of even quite modest size are able to form at near-collisional rates. The radiative association of radicals CH_3 and CN to form methyl cyanide

$$CH_3 + CN \rightarrow CH_3CN + h\nu$$

is an example of this type of mechanism; the rate coefficient for this reaction is estimated to be about 10^{-16} cm^3 s^{-1} for temperatures in the range 10–300 K. Therefore, this association, involving half a dozen atoms, is about a hundred times more efficient than the radiative association of H and H^+. The rate coefficient for these reactions is highly sensitive to the number of atoms in the transient complex, and for quite modest numbers the reaction may proceed with almost a collisional rate coefficient, typically $\sim 10^{-11}$ cm^3 s^{-1}.

3.2.2 Exchange Reactions

Neutral Exchange Reactions
If some molecules are already present, then exchange reactions may be able to convert one molecule into another. Assuming that the reaction between the neutral atom A and neutral molecule BC

$$A + BC \rightarrow AB + C$$

can proceed (*i.e.*, is energetically favourable, or exothermic), the exchange converts the molecule BC into a new molecule AB. Since we know that molecular hydrogen is very abundant in some interstellar and circumstellar clouds (see Figure 2.2), then this type of reaction may in principle be capable of converting some of the H_2 molecules into interstellar hydrides such as OH, CH or NH. However, exchange reactions of O, C or N atoms with H_2 to form hydrides are endothermic and can only contribute significantly at high temperatures.

Exothermic exchange reactions between neutral partners are often found to be impeded by an energy barrier which may suppress the reaction at low temperatures such as those of cold interstellar clouds; however, the reaction may proceed at about the collisional rate at more elevated temperatures. We may think of how the barrier is formed by imagining a contour diagram that plots the interaction of the three atoms, A, B, and C, as a surface whose position depends on the distances between the three atoms. When A is at large distances from B and C, the diagram shows the potential well of the molecule BC as a valley in the contour surface, while when C is at large distances from A and B, the diagram shows the potential well of the molecule AB as a valley in the contour surface, see Figure 3.1. The barrier may occur

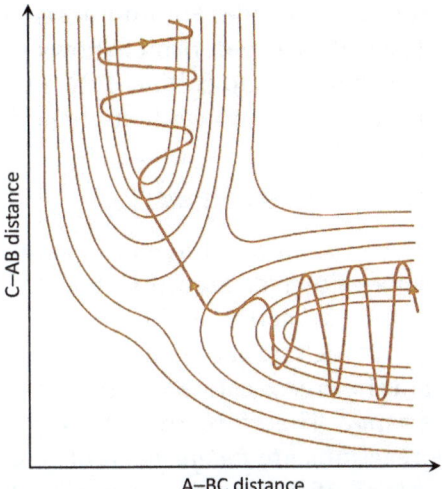

Figure 3.1 The interaction of A, B, and C represented schematically as a potential energy contour map. At large separations, oscillations in the two valleys represent the vibrating molecules AB and BC. Energy is required to pass from the head of one valley to the other. This is the barrier to the reaction AB + C → A + BC. Quantum mechanical tunnelling through the barrier may occur.

where the two valleys join. If there is sufficient energy in A–BC to surmount the barrier, the system may emerge as AB–C, so a conversion will have taken place.

If a barrier exists, then there the rate coefficient of the reaction has a strong temperature dependence. For example, the reaction of atomic nitrogen with molecular hydrogen (where the barrier is about an electron volt) to form nitrogen hydride

$$N + H_2 \rightarrow NH + H$$

has a rate coefficient of about $2.7 \times 10^{-10} \exp(-12\,600/T)$ cm^3 s^{-1} for temperatures of two to three thousand K.

Ion-molecule Exchange Reactions
This picture is changed entirely if the incident atom is ionized. Many exothermic reactions of the type

$$A^+ + BC \rightarrow AB^+ + C$$

(*ion–molecule exchange reactions*) are known to occur without activation and at roughly the collisional rate. The ion induces a dipole in the molecule and the ion–dipole interaction can be shown to be $\alpha e^2/(8\pi\varepsilon_0 r^4)$, where α is the polarizability of the molecule, e is the electronic charge, ε_0 is the permittivity of free space, and r is the distance between the ion and the molecule. For an interaction energy varying with r as r^{-4}, the classical theory of orbits shows that the interacting pair will spiral into each other and collide if the *impact parameter* (the shortest distance between the molecule and the initial direction of the ion) is less than b_0, where

$$b_0 = (\alpha e^2 / \pi \varepsilon_0 \mu v^2)^{1/4},$$

μ is the reduced mass $[\mu = m_A m_{BC}/(m_A + m_{BC})]$, and v is the velocity of approach at large separation (see Figure 3.2).

The cross section for interaction is therefore πb_0^2 (note that this is proportional to $1/v$) and the rate coefficient for the reaction is the integral of the product of the cross section multiplied by velocity, over the whole temperature range. Therefore, the rate coefficients for exothermic ion–molecule reactions are *independent* of velocity, a remarkable result. In fact, the rate coefficients depend only on the polarizability of the initial molecule and on the reduced mass; however, these parameters do not vary greatly for many of the likely partners, so many of the measured rate coefficients for exothermic ion–molecule reactions should have very roughly similar values (as is found to be the case). These reactions are fairly energetic, releasing in the final collision enough energy (about an electron volt) to overcome internal energy barriers to rearrangement.

An example of such an exothermic ion–molecule reaction is the *H-abstraction reaction*, a type of reaction that occurs frequently in astrochemistry because hydrogen molecules are relatively so abundant; for example:

$$O^+ + H_2 \rightarrow OH^+ + H$$

This reaction is fast (the rate coefficient is about 10^{-9} cm^3 s^{-1}) and is independent of temperature for temperatures less than about 300 K.

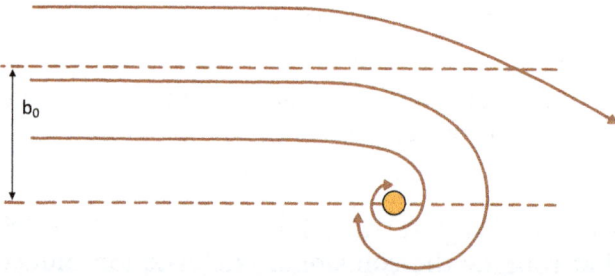

Figure 3.2   Trajectories in an r^{-4} potential between a charge and the induced dipole. If the impact parameter is less than a critical value b_0, then the orbital path is a spiral and interaction occurs.

Note that the corresponding reaction with C^+ is endothermic by about an energy equivalent to 4640 K, so it does not occur in cold interstellar clouds (but can be important in warm regions).

The OH^+ ion, however, may undergo further H-abstraction reactions:

$$OH^+ + H_2 \rightarrow OH_2^+ + H$$

and

$$OH_2^+ + H_2 \rightarrow H_3O^+ + H$$

and recombinations of these products with electrons lead to the formation of neutral products, OH and H_2O, both of which are important astrochemical species:

$$OH_2^+ + e \rightarrow OH + H$$

and

$$H_3O^+ + e \rightarrow H_2O + H$$

These processes are called *dissociative recombinations*, and are very efficient, as we shall discuss in more detail in Section 3.2.3.

Helium Ion Reactions

Finally, in our discussion of ion–molecule reactions, it is worth considering the particular case of reactions of molecules with singly ionized helium, He^+. This ion has a significant role in interstellar chemistry and is formed by cosmic ray (c.r.) impact on the neutral atom:

$$He + c.r. \rightarrow He^+ + c.r. + e$$

These helium ions do *not* lead to a large range of exotic helium-containing molecules. In fact, these ions are deadly for molecules. Helium has a very large ionization potential of 24.6 eV, and so the helium ion grabs an electron at any opportunity, such as a collision. For example, the ionization potential of CO is significantly smaller than that of helium, at ~14 eV. In a charge exchange reaction of He^+ with CO, so much energy remains in the collision complex that the carbon monoxide dissociates:

$$CO + He^+ \rightarrow C^+ + O + He$$

This destructive type of reaction is called *dissociative charge exchange*. In fact, the helium ion ionizes or dissociates almost any other molecule in the interstellar medium by this kind of reaction. These reactions are important constraints on the chemical network. It's worth noting, however, that a helium-bearing molecule, HeH^+, was important in the chemistry in the early Universe (see Chapter 5), and has recently been detected by means of emission in its 1-0 rotational transition in a planetary nebula. In molecular gas, HeH^+ is formed in the reaction of HCO and He^+, while in HII regions and in the early Universe (see Section 5.7.1) it is formed through the radiative association of ionized hydrogen and neutral helium, as well as the radiative association of neutral hydrogen and ionized helium.

3.2.3 Recombination of Atomic and Molecular Ions with Electrons

Radiative Recombination
In the process of recombination on to an atomic ion, X^+,

$$X^+ + e \rightarrow X + h\nu$$

the electron is captured into a specific state of X^+ that was initially unoccupied, and a photon is emitted that has energy of the sum of the initial energy of the electron and the ionization potential of the state into which capture occurs. Figure 3.3 illustrates the process: *radiative recombination*.

The rate coefficients for radiative recombination on to atomic ions may be calculated accurately. Radiative recombination is relatively slow, because the photon must be emitted in the short interval in which the electron is in the vicinity of the ion. Hence, the rate coefficients for radiative recombination are slightly larger at lower temperatures. For example, for the radiative recombination of singly ionized carbon atoms with electrons, the rate coefficient is fairly small: $4.4 \times 10^{-12} (T/300)^{-0.61}$ cm^3 s^{-1}.

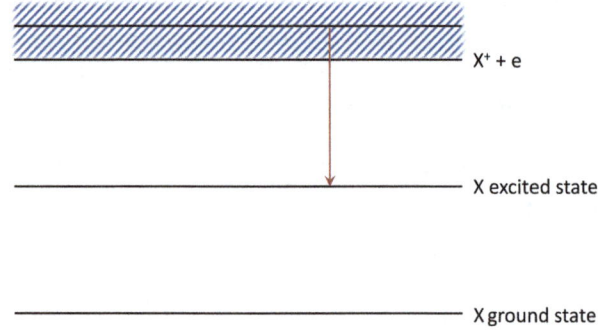

Figure 3.3 Radiative recombination occurs from the continuum of X^+ into a bound state of X. Schematic ground state, excited state, and ionization energy levels of atom X are shown. The shaded region indicates the ionization continuum. Recombination can occur from the continuum to a bound level of X, as indicated. A photon of corresponding energy is emitted.

Dielectronic Recombination

There is an additional recombination process for ions that have at least one bound electron; this process involves the incident electron and the bound electron, and is called dielectronic recombination. The incident electron may transfer energy to a bound electron raising it into an excited state of the atom, while removing enough energy from the incident electron so that it may be captured into an excited state. Subsequently, one or both of the two electrons may spontaneously relax into a lower energy state so that the atom no longer has enough energy to reverse the process, *i.e.*, it cannot *autoionize*.

Obviously, this process cannot operate for hydrogen ions as they do not have a bound electron, but for other ions the dielectronic recombination coefficient can be several times larger than that for radiative recombination. For example, radiative and dielectronic recombination coefficients for C^+, C_2^+, and C_3^+ at 10^4 K are (4.66 and 1.84), (24.5 and 60.6), and (50.5 and 131), respectively, in units of 10^{-13} cm^3 s^{-1}.

Dissociative Recombination

This process occurs in the recombination of electrons with *molecular* ions

$$AB^+ + e \rightarrow AB^* \rightarrow A + B$$

It is very efficient and therefore an important mechanism for the removal of ions and electrons in interstellar chemical networks. The recombination effectively deposits energy equivalent to the ionization potential of AB into the molecule. This is generally a much larger amount of energy than the dissociation energy of AB, and so the atoms of excited molecule AB* will begin to separate, and this occurs on a timescale comparable with a vibration period, typically $\sim 10^{-13}$ s. The alternative relaxation for AB* is emission of a photon. However, even the shortest timescales for spontaneous emission of radiation are much longer, typically $\geq 10^{-8}$ s, so AB* almost always relaxes by separation with kinetic energy equal to the ionization potential of AB (typically ~ 10 eV), less the binding energy of AB (typically ~ 5 eV). Consequently, the rate coefficients for dissociative recombinations are often very large. For example, the rate coefficient for dissociative recombination

$$CH_2^+ + e \rightarrow C + H + H$$

is $4.0 \times 10^{-7}(T/300)^{-0.6}$ cm^3 s^{-1}, and there are also other fast dissociation channels (C+ H_2 and H + CH). This is five orders of magnitude larger than the rate coefficient for radiative recombination for carbon ions with electrons, quoted above.

We have seen (Section 3.2.2) in our discussion of ion–molecule reactions that in interstellar clouds these processes lead efficiently to molecular ions such as OH^+ and that hydrogen abstraction reactions may create hydrogenated molecular ions such as H_2O^+ and H_3O^+. All of these molecular ions are subject to rapid dissociative recombination reactions, so at least some of these species lead to neutral molecules (OH and H_2O, in this example). These routes are of great importance in gas-phase ion–molecule chemistry, particularly in the interstellar medium, as we shall see in Chapter 4.

In fact, gas phase chemistry in low temperature interstellar clouds depends to a large extent on ion–molecule chemistry. The observed preponderance of neutral molecules (rather than ions) is accounted for by efficient recombinations between molecular ions and electrons. So the main question to address is this: since interstellar ionization may drive much of interstellar chemistry in cold interstellar gas clouds, what causes the ionization in these regions? Obviously, the ionization needs to be continuously replenished, because each ion–molecule chemical reaction sequence ends in a recombination.

3.2.4 Photoionization of Interstellar Atoms by Starlight

By the standards of the Milky Way, the Sun is a star of modest mass with effective temperature about 5800 K and a relatively feeble intensity in a spectrum that stretches from the near ultraviolet (wavelength around 250 nm), peaks at around 500 nm (in the green part of the visual spectrum) and extends to near infrared wavelengths of a few μm. Radiation from solar-type stars is clearly not a significant source of ionization in the interstellar medium. However, there are some stars in the Galaxy that are very much hotter than the Sun, and they have a dramatically different spectrum. A star with effective temperature of, say, 30 000 K has an intensity that is hundreds of times brighter than the Sun and a spectrum that peaks at much shorter wavelengths than the visible region, in fact in the far ultraviolet Figure 3.4 compares black body energy emission rates (as a rough approximation for stellar emissions) for several different black body temperatures.

Radiation of wavelengths shorter than 91.2 nm is energetic enough to photoionize atomic hydrogen, the most abundant interstellar species (since atomic hydrogen has an ionization potential of 13.6 eV). For a hot star embedded in atomic hydrogen, each photoionization is balanced by a recombination, so the ionizing radiation weakens with distance from the star. Observations (see Figure 1.2) supported

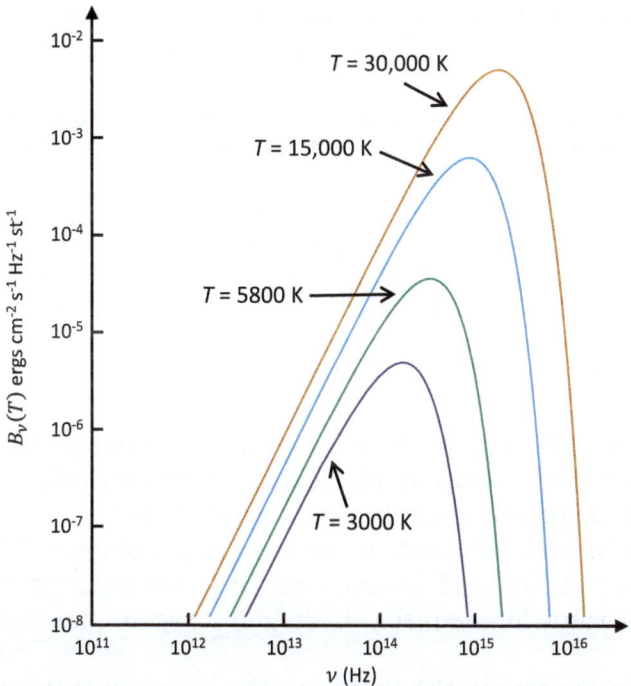

Figure 3.4   Energy emission rates from black bodies of several different temperatures. The black body temperature of 3000 K peaks in the near infrared, while that at 5800 K (corresponding to the Sun) peaks in the visible. The temperature of 30 000 K may correspond to emission from a very hot star; its emission intensity is very much higher than for black bodies at lower temperatures and peaks in the ultraviolet.

by theoretical study show that the transition between almost fully ionized hydrogen (HII in spectroscopic notation) to almost fully neutral hydrogen (HI) is very narrow. Therefore, these hot stars are embedded in near-spherical regions of almost fully ionized hydrogen (HII regions) and that largely neutral atomic hydrogen is outside these regions. Recombinations of protons and electrons within these spheres occur *via* transitions between excited states of atomic hydrogen. Transitions between highly excited states of atomic hydrogen generate a radio spectrum. Many of the recombinations eventually involve transitions between principal quantum numbers $3 \rightarrow 2$, which generate the intense visible red emission at 656 nm (so-called Hα emission in the Balmer series), prominent in many astronomical images (see Figure 1.2).

A crucial implication for interstellar chemistry of the structure of HII regions is that all stellar ultraviolet photons with energies ≥13.6 eV are trapped in the HII regions, so outside those regions all photons have energies less than 13.6 eV. This has several consequences:

- Since HII regions are flooded with very high energy photons of starlight and are therefore hostile to molecule formation, abundant molecules can exist only *outside* HII regions.
- A chemical network operating outside HII regions needs to take into account photodissociation or photoionization of atoms and molecules only by photons of starlight with energies less than 13.6 eV.
- Atoms in the interstellar medium outside HII regions may be either neutral or ionized, depending on whether their ionization potentials are greater or less than 13.6 eV (see Table 3.1).

It is clear that in unshielded regions of the interstellar medium outside HII regions, some of the elements in atomic form in the interstellar gas will be photoionized by the radiation fields escaping from HII regions. These include carbon, sodium, magnesium, silicon, sulfur, potassium, calcium and its ion, and iron. However, three elements most important in chemistry, hydrogen, oxygen, and nitrogen, are neutral.

Hot stars are rare in the Galaxy, so the intensity of the interstellar radiation field varies from place to place in interstellar space. Also, there is a range of effective temperatures of the hot stars, so their spectra differ to some extent. Measurements and theoretical estimates of the interstellar radiation field agree reasonably well, so that for the purposes of interstellar chemistry, a mean intensity and mean spectrum of the interstellar radiation field in the Milky Way galaxy can be adopted, and rates of photodissociation and photoionization can be calculated with respect to that mean field, unshielded by interstellar dust grains. These results are included in compilations of astrochemical data. We shall discuss these databases in Chapter 5. The adopted radiation field energy density is fairly flat through the visible and near ultraviolet with a value in the approximate range $(3–7) \times 10^{-21}$ J cm^{-3}

Table 3.1 Ionization potentials (I.P. in eV) of some species of astrochemical interest.

Atom	I.P.	Atom	I.P.	Atom	I.P.	Atom	I. P.	Molecule	I.P.
H	13.60	O	13.62	Si	8.15	K$^+$	31.63	H$_2$	15.43
He	24.59	Na	5.14	Si$^+$	16.35	Ca	6.11	C$_2$	12.00
C	11.26	Na$^+$	47.29	S	10.36	Ca$^+$	11.87	CN	13.80
C$^+$	24.38	Mg	7.65	S$^+$	23.34	Fe	7.90	CO	14.01
N	14.53	Mg$^+$	15.04	K	4.34	Fe$^+$	16.19	CH$_4$	12.60

and (of course) a cut-off to zero at 91.2 nm (13.6 eV). Of course, in modelling the chemistry in interstellar clouds, allowance must be made for the effects of interstellar extinction. The databases show how this can be done.

3.2.5 Photodissociation and Photoionization of Molecules by the Interstellar Radiation Field

Molecules may be photoionized or photodissociated by the interstellar radiation field. Figure 3.5 indicates schematically how (a) photoionization and (b) photodissociation arise in a conventional way for a diatomic molecule.

In Figure 3.5(a), a transition occurs from a vibrational state of molecule AB (separated state A + B) to a vibrational state of molecule AB^+ + e (separated state A^+ + B + e). This is photoionization. In Figure 3.5(b), a transition occurs from a vibrational state of AB (separated state A + B) to a repulsive state (separated state also A + B). This is photodissociation. Rates of photodissociation for simple molecules irradiated by the mean unshielded interstellar radiation field can be as large as $\sim 10^{-9}\,s^{-1}$ and in diffuse clouds with low extinction this is very often the fastest loss mechanism.

Photodissociation can involve several exit channels. For example, CH_4, may be photodissociated in various ways:

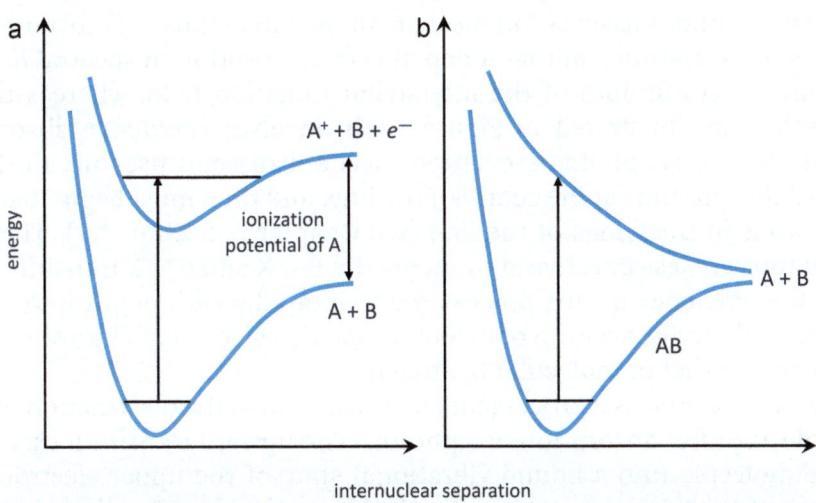

Figure 3.5 Potential energy curves and transitions for (a) photoionization (on the left) and (b) photodissociation (on the right) in a diatomic molecule.

$$CH_4 + h\nu \rightarrow H + CH + H_2$$
$$\rightarrow H_2 + CH_2$$
$$\rightarrow H + CH_3$$

in addition to photoionization:

$$\rightarrow CH_4^+ + e.$$

It is clear from Table 3.1 that the most abundant molecule in the Galaxy, H_2, cannot be photoionized by the interstellar radiation field available outside HII regions ($h\nu < 13.6$ eV), and a direct photodissociation of H_2 also requires photons of energies not available in the diffuse clouds. Does this mean that molecular hydrogen is stable against the interstellar radiation field? No. Molecular hydrogen can be photodissociated by this interstellar radiation field using photons of energy that are still available in the interstellar medium, of about 12 eV (wavelength ~ 100 nm), in a subtle indirect mechanism. The situation is illustrated in Figure 3.6. Line absorption at about 100 nm in the Lyman and Werner bands of H_2 from the ground electronic state, $X^1\Sigma_g^+$, and the lowest vibrational level, $v'' = 0$, populates the various rovibrational levels of the first and second excited electronic states $B^1\Sigma_u^+$ and $C^1\Pi_u$. The excited molecules relax almost immediately into the ground electronic state, but into a variety of vibrational levels, including the levels in the vibrational continuum. The fraction of excitations falling into continuum levels is the fraction of the excitations leading to immediate H_2 dissociation.

The H_2 photodissociation mechanism is unlike those of most other photodissociations, in that it depends on absorption in spectral *lines* from the continuum of the interstellar radiation field, whereas the mechanism illustrated in Figure 3.6(b) involves *continuum* absorption. So the H_2 photodissociation mechanism soon uses up all the available photons at the centre of the line, and then must begin to use photons in the wings of the line (see Chapter 2, Section 2.1). These photons are less effective at inducing the B ← X and C ← X transitions, so the efficiency of this process reduces rapidly with depth into the cloud. Photodissociation of carbon monoxide also occurs by a similar process to that of molecular hydrogen.

Predissociation is a process that may also lead to the dissociation of a molecule after absorption of a photon. The upward transition excites the molecule into a bound vibrational state of the upper electronic state. If this upper state mixes with the continuum of a dissociative state of the molecule, then the molecule may dissociate. Predissociation therefore occurs in discrete lines corresponding to state-to-state vibrational transitions.

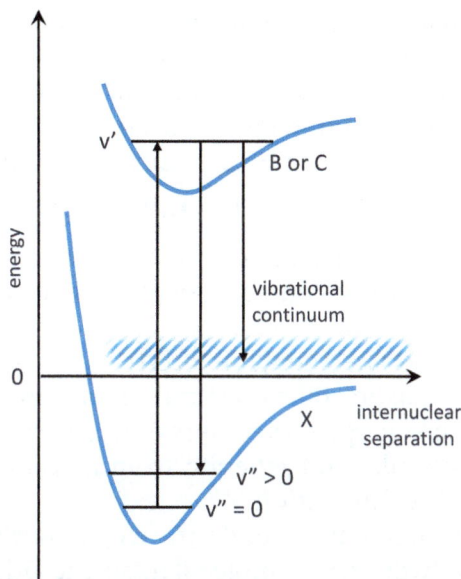

Figure 3.6 A schematic representation of the potential energy curves X and B or C of the H_2 molecule. Molecules can be excited in some transitions from ground state X to excited states B or C by photons that are available in the interstellar gas; the molecules rapidly relax from states B or C, arriving in either the vibrational bound or the vibrational continuum (shaded area) states of the ground electronic state X. In the vibrational continuum, the molecules "fall apart" with an average kinetic energy of about 0.4 eV in the dissociating H atoms. Most of the transitions (~80%) from states B and C populated by the interstellar radiation field end in bound vibrational levels of the ground state X. Transitions between bound vibrational levels of X are slow, so a significant population of excited vibrational levels in the ground state X is maintained.

3.2.6 Collisional Ionization by Cosmic Rays, and Generation of the Cosmic Ray-induced Radiation Field

We have seen that radiation from hot stars causes ionization in interstellar gas, and that this ionization can enable ion–molecule chemistry to occur. However, starlight is excluded from dense interstellar clouds because of the extinction caused by interstellar dust grains mixed with the gas. However, ionization is also created by cosmic rays (described in Chapter 1, Section 1.1.3), and these energetic particles are capable of penetrating very dense and dusty interstellar gas. The cosmic rays are mainly high energy protons. The most abundant cosmic rays that can be detected have energies of a few MeV and these protons at the low-energy end of the cosmic ray range are the most effective in ionizing any interstellar atoms and molecules. Since hydrogen atoms and molecules are the most abundant species in the interstellar medium, the ions H^+ and H_2^+ created by cosmic ray ionization:

$$(H, H_2) + c.r. \rightarrow (H^+, H_2^+) + e + c.r.$$

are very important in interstellar chemistry. The cosmic rays involved in these ionizations are, of course, negligibly affected by this process, but the electron carries off about 30 eV if the cosmic rays have energies of about 2 MeV. The energetic electrons can share energy with other particles in elastic collisions, and are an important source of heat in the gas. The energetic electrons may also excite transitions in H atoms or H_2 molecules and generate an emission spectrum, as we'll see at the end of this sub-section.

In dense dark interstellar clouds in which starlight is excluded, almost all of the hydrogen is molecular, so the primary ion is H_2^+, which rapidly undergoes a hydrogen abstraction with H_2:

$$H_2^+ + H_2 \rightarrow H_3^+ + H$$

This new molecular ion turns out to be an important driver of ion–molecule chemistry in dense dark interstellar clouds because the binding (or *proton affinity*) of H^+ to H_2 in H_3^+ is weaker than the proton affinity in many other molecules. Table 3.2 shows that H_3^+ has a lower proton affinity than many other molecules of interest, and will transfer a proton very readily.

The result of a proton transfer may open up chemical routes to other species. For example, transferring a proton to methanol

$$CH_3OH + H_3^+ \rightarrow CH_3OH_2^+ + H_2$$

Table 3.2 Proton affinities for some molecules of astrochemical interest. Molecular hydrogen has a lower proton affinity than all but one of the molecules listed, so H_3^+ can transfer a proton to any of these molecules, other than O_2.

Molecule	Proton affinity (eV)	Molecule	Proton affinity (eV)
H_2	4.42	CH_4	5.57
CO	6.19	H_2O	7.20
N_2	5.15	NH_3	8.90
NO	5.53	H_2CO	7.43
O_2	4.38	HCN	7.43
SH	7.19	CH_3OH	7.85
CS	7.63	CO_2	5.64

may even open a pathway from methanol to dimethyl ether

$$CH_3OH_2^+ + CH_3OH \rightarrow CH_3OCH_4^+ + H_2O$$

$$CH_3OCH_4^+ + e \rightarrow CH_3OCH_3 + H$$

or other paths to chemical complexity.

As stated earlier, the energetic electron released in the cosmic ray ionization of H_2 may share some of its energy in elastic collisions with other species, helping to heat the gas. It may also excite transitions in H_2; these may be electronic, vibrational, or rotational transitions. Electronic transitions from excited electronic states to the ground state, X, occur rapidly, radiating in the ultraviolet. This creates a local ultraviolet radiation deep inside a dark cloud, called the *cosmic ray induced radiation field*. It is quite independent of the interstellar radiation field external to the cloud which arises from ambient starlight from massive stars. However, the cosmic ray induced radiation field is weak; its intensity is only about 10^{-4} of that of the typical unshielded external interstellar radiation field created by starlight. Nevertheless, it is independent of extinction by interstellar dust and has an effect on interstellar chemistry inside dark clouds.

3.2.7 Ionization by X-rays

Cosmic X-rays are emitted by energetic objects such as active galaxies or massive stars (see, *e.g.*, Figure 6.5b), and also from the very high temperature coronal gas that occupies much of interstellar space. X-rays are also emitted by modest stars like the Sun, and the young Sun was very much brighter in X-rays than it is at present, so may have had an effect on molecules during the early evolution of the solar system.

The effect that X-rays have on interstellar matter is that they can ionize all atoms and molecules. Near a very powerful X-ray source, they will do so and the subsequent dissociative recombinations will therefore tend to suppress all interstellar chemistry. However, X-rays of the shortest wavelengths (<1 nm) interact only weakly with atoms and molecules, so they are only capable of causing a low level of ionization – but this will affect a very large volume of space. Every atom and molecule is capable of being ionized, just as in the case of cosmic ray ionization, but unlike the case of ionization by starlight (see Section 3.2.5). The most important effects of the low-level X-ray ionization are in dark clouds, and since the most abundant species by far is

H_2, the effects of X-ray ionization appear through the formation of the H_3^+ ion:

$$H_2 + \text{X-ray} \rightarrow H_2^+ + e$$

$$H_2^+ + H_2 \rightarrow H_3^+ + H$$

followed by reactions with O and C atoms, for example, to provide the reactive ions H_3O^+ and CH_3^+, which are very important in developing chemical complexity in interstellar chemistry (see Chapter 4).

3.2.8 Interstellar Anions

Detected molecules (see the Appendix to Chapter 1) are mostly neutral species. Some positive molecular ions (*i.e.*, cations) are also detected. They are formed in reactions with positive ions formed by photoionization and collisional ionization. A number of interstellar *anions* (including, for example, C_3N^-, C_4H^-, and C_6H^-) have also been detected. It appears that the large electron affinities of carbon chains and hydrocarbons with more than 4 or 5 atoms allow the efficient formation of negative ions through the process of *radiative electron attachment:*

$$X + e \rightarrow X^- + h\nu$$

where X may be a neutral atom or a neutral molecule. For example, the radiative attachment of C_4H with electrons has a large rate coefficient of about $1.1 \times 10^{-8}(T/300 \text{ K})^{-0.5}$ cm^3 s^{-1}, implying an efficient route to the anion C_4H^-.

Once negative ions are formed, they are subject to a number of destruction mechanisms, such as *mutual neutralization:*

$$A^- + B^+ \rightarrow A + B,$$

which is obviously very likely to be efficient because of the strong attraction between the charges. *Associative detachment* also occurs:

$$A^- + B \rightarrow AB + e$$

and is also an efficient process, and *photodetachment*

$$A^- + h\nu \rightarrow A + e$$

is important in regions where there is a significant radiation field. Loss of anions also occurs in *collisions with neutral atoms,* leading to a variety of possible products.

The anion H$^-$ is of interest, as it provides a possible route to the formation of molecular hydrogen in the associative detachment reaction:

$$H^- + H \rightarrow H_2 + e$$

However, as we shall discuss further in Chapter 7, this route cannot provide H_2 at a high enough rate to account for its high abundance in the Milky Way galaxy.

3.2.9 Chemistry in Interstellar Regions Heated by Shocks

Interstellar regions are not static but dynamic. Gas flows set in motion by stellar explosions or stellar winds impinge on other gas in the vicinity, such as interstellar clouds. The flows may therefore be arrested by shocks in which the kinetic energy of gas flows is converted to thermal energy when those winds collide with interstellar clouds. The temperature rise can be very large indeed. Shocks from supernovae explosions, injecting gas at speeds of thousands of kilometres per second into interstellar space and colliding with ambient gas, may raise the temperature of the ejecta and the interstellar gas to millions of degrees K and create the so-called coronal gas that occupies much of interstellar space (see Table 1.3a).

However, interstellar clouds are often seen to be in relative motion to one another with much lower velocities, typically a few tens of kilometres per second. Collisions between interstellar clouds at these much lower velocities cause more modest but still significant abrupt and rather short-lived temperature rises to some thousands of kelvin. Molecules not only survive at these temperatures but also may be formed in reactions that are inhibited by energy barriers at the normal low temperatures of interstellar molecular clouds (~10 K). As a result, direct reactions of oxygen atoms with H_2 molecules at temperatures above a thousand degrees produce hydroxyl and water molecules, and similar reactions with C^+ ions produce CH^+ molecules. Nitrogen atoms form nitrogen hydrides, NH and NH_2, but the final step to ammonia is slow. In diffuse clouds, sulfur is present as S^+, which reacts successively with H_2 in warm shocked gas to form H_3S^+, and this ion dissociatively recombines with electrons to give SH and H_2S. In exchange reactions with oxygen, these molecules then give SO and SO_2. In dark clouds, sulfur is present as neutral atoms which in warm gas react directly with H_2 to give sulfur hydrides which then form the oxides in neutral exchanges with O atoms and OH molecules.

However, this fertile shock chemistry is very sensitive to shock speed. For shock speed values above about 25 km s^{-1} (or 45 km s^{-1} if the shock is modulated by a magnetic field), molecular hydrogen itself is destroyed in the shock, so the gas is then composed mainly of hot atomic hydrogen (the post shock temperature may approach 10^4 K). Hot atomic hydrogen is an extremely destructive species, even for very strongly bound molecules such as CO, in reactions such as

$$H^* + CO \rightarrow C + OH$$

$$H^* + OH \rightarrow H_2 + O$$

$$H^* + H_2 \rightarrow 3H$$

where H* is a H atom with a large amount of thermal energy. All interstellar molecules are very quickly destroyed by hot atomic hydrogen.

3.2.10   Isotope Chemistry and Isotopologues

In the Appendix to Chapter 1, detected species are shown only for the main isotopologues. But as remarked there, many molecules are detected in a variety of isotopologues, so the number of detected species is in fact much larger than shown. Detected isotopologues in interstellar gas often show a remarkable abundance enhancement of what should be a minor species. To be specific, the average ratio of deuterium to hydrogen in the Galaxy is about 1.6×10^{-5}, but the abundance ratio for the isotopologues containing D and H is always much larger. For example, the abundance ratio for $HDCO : H_2CO$ is 0.14, and for $CH_2DOH : CH_3OH$, it is 0.03. How can this deuterium enhancement (called *isotopic fractionation*) by such large factors ($\sim 10^4$ in the example of formaldehyde) be achieved?

The answer lies in a small difference between the energy levels of the key species in ion–molecule chemistry, H_3^+ and H_2D^+. The larger mass of D than H means that the energy levels of H_2D^+ lie slightly lower than those of H_3^+. Hence, the reaction

$$H_3^+ + HD \leftrightarrow H_2D^+ + H_2 + 227 \text{ K}$$

occurs preferentially in the forward direction for all temperatures less than 227 K. This means that interstellar protonated molecular hydrogen contains more D than H in gas at low temperatures ($T < 227$ K) than would be the case at higher temperatures ($T > 227$ K) when the rates of forward and reverse reactions are in balance. At $T \sim 10$ K, D is transferred preferentially into H_2D^+ and in subsequent reactions such as those involving H_2CO and CH_3OH, at sufficiently low temperatures. Similar fractionation occurs for many other species.

3.2.11 Three-body Reactions

As mentioned in Section 3.2.1, the collision of a third body, M, with a colliding pair, AB*, is capable of removing energy from the colliding pair, so stabilizing the pair and forming the bound molecule, AB:

$$A + B + M \rightarrow AB^* + M \rightarrow AB + M^*$$

where M* is M with kinetic (or other) excitation. Since AB* is very short-lived, the stabilizing collision must occur very rapidly indeed, and this implies that very high gas number densities (compared to the normal range of interstellar densities) are required to make this process competitive with the two-body processes described in the previous sections of this chapter. Rate coefficients for the three-body reaction may be crudely estimated to be very roughly $\sim 10^{-32}$ cm^6 s^{-1} and this suggests that number densities at least on the order of 10^{10} cm^{-3} are required for three-body reactions to be important. Number densities of this magnitude are met in stellar atmospheres and envelopes, and in stellar explosions, and therefore chemistry in such regions may include three-body reactions. The chemistry of these regions will be described in Chapter 6. Of course, planetary atmospheres (including Earth's atmosphere, where the number density is $>10^{19}$ cm^{-3}) may have number densities vastly in excess of 10^{10} cm^{-3} so their chemistries are dominated by three-body reactions.

Rate coefficients, k_3, for various three-body reactions have been determined, and a few examples are given here in which molecular nitrogen is the third body (taken from the KIDA database[2]):

$$H + H + N_2 \rightarrow H_2 + N_2; \quad k_3 = 9.14 \times 10^{-33}(T/300)^{-0.6} \text{ cm}^6 \text{ s}^{-1}$$

$$C + H_2 + N_2 \rightarrow CH_2 + N_2; \quad k_3 = 7.00 \times 10^{-32} \text{ cm}^6 \text{ s}^{-1}$$

$$C + C + N_2 \rightarrow C_2 + N_2; \quad k_3 = 5.30 \times 10^{-31}(T/300)^{-1.6} \text{ cm}^6 \text{ s}^{-1}$$

In general, the formation rate of AB in a three-body reaction (see the start of this sub-section) is $k_3\, n(A)n(B)n(M)$ cm^{-3} s^{-1}.

3.3 Photon-driven Chemistry in Astronomy

There are two specific terms in occasional use in discussions of photon-driven chemistry in astronomical situations: these are *photochemistry* and *radiation chemistry*. These terms have specific definitions, as follows:[3]

Photochemistry: chemistry in which chemical processes are initiated by photon-induced excitations that do not involve ionization;

Radiation chemistry: chemistry in which chemical processes are initiated by photon-induced excitations of sufficiently great energy to generate ionization.

The chemistry of interstellar clouds is essentially driven by a combination of several possible contributions: (i) from interactions of atoms and molecules with the interstellar radiation field (see Section 3.2.4); (ii) from ionization of atoms and molecules by cosmic rays (see Section 1.1.3); and, as we'll discuss in Chapters 7 and 8, (iii) from processes on interstellar dust grains. As we shall describe in more detail in Chapters 4 and 5, interactions with the interstellar radiation field are significant in driving chemistry in diffuse clouds but are relatively unimportant in dark clouds in which starlight is excluded by extinction caused by interstellar dust (Section 1.1.4). We have seen (in Section 3.2.4) that the interstellar radiation field in diffuse clouds is capable of ionizing (*i.e.*, *radiation chemistry*) some atomic species important in chemistry such as carbon, silicon, and sulfur, but cannot ionize atoms of hydrogen, oxygen, and nitrogen. The radiation field in diffuse clouds is also important in photoionizing (*i.e.*, *radiation chemistry*) and photodissociating (*i.e.*, *photochemistry*) some simple molecules.

There are near-stellar situations in which gases may be subjected to radiation fields of wavelengths much shorter (and more energetic) than those of the interstellar radiation field. As mentioned earlier, young stars similar to the Sun are sources of X-ray emission. This radiation will photoionize atoms and molecules, ejecting primary photoelectrons of sufficient energy to generate secondary electrons which cause further ionization; and ultimately they cause heating of the gas. This is an example in which *radiation chemistry* has a powerful role.[4] X-rays may also drive *radiation chemistry* in material in the gas phase or in ices.

3.4   Summary

In this chapter we have discussed the main types of reaction that contribute significantly to gas phase astrochemistry. In the next chapter, Chapter 4, we shall describe the chemical networks that are believed to generate the molecules that are detected in various regions of space. Evidently, the data requirements for this work are enormous. These requirements have been met by fundamental theoretical studies and by dedicated laboratory investigations. We shall describe in Chapter 5 some of the databases that have been generated by this work and we shall show how these databases are used in modern astrochemical applications. It is worth emphasising here that while gas phase processes are essential and a major component of astrochemistry, they are only one part of the picture, and that chemistry occurring on the surfaces of dust and in the ices that may be deposited on those surfaces is also necessary for a proper understanding of astrochemistry. We shall discuss those aspects of astrochemistry in Chapters 6–8.

References

1. A. G. G. M. Tielens, *Molecular Astrophysics*, Cambridge University Press, 2021.
2. *KIDA: Kinetic Database for Astrochemistry*: http://kida.astrophy.u-bordeaux.fr.
3. C. R. Arumainayagam, R. T. Garrod, M. C. Boyer, A. K. Hay, S. T. Bao, J. S. Campbell, J. Wang, C. M. Nowak, M. R. Arumainayagam and P. J. Hodge, *Chem. Soc. Rev.*, 2019, **48**, 2293.
4. D. Locci, A. Petralia, G. Micela, A. Maggio, A. Ciaravella and C. Cecchi-Pestellini, *Planet. Sci. J.*, 2022, **3**, 1.

4 Gas Phase Chemical Networks in Interstellar Clouds

"Models of interstellar clouds should be able to account for the amounts of atomic and molecular hydrogen, the abundances of simple molecules, the fractionation of HD, the degree of ionization of minor atomic species, and the populations of the rotational levels of the ground state of H_2. If a model can be constructed that reproduces the array of observational data, some confidence may be placed in the inferred physical parameters, such as the density, the temperature, the strengths of the ultraviolet field, and the intensity of the ionization source." (J H Black and A Dalgarno)[1]

4.1 Introduction

In Chapter 3 we introduced many types of gas-phase reaction, particular examples of which may contribute to the formation of molecules in interstellar and circumstellar space. We described the physical situations in which these reactions may be important. In this chapter we shall identify the reaction schemes, or networks, in which chemistry of various elements is believed to be initiated in low temperature interstellar clouds. These first steps in gas-phase astrochemistry often seem to be the most difficult, but once they have been taken,

Astrochemistry: Chemistry in Interstellar and Circumstellar Space
By David A. Williams and Cesare Cecchi-Pestellini
© David A. Williams and Cesare Cecchi-Pestellini 2023
Published by the Royal Society of Chemistry, www.rsc.org

increasing the range of molecular species and achieving greater complexity is often straightforward. We shall consider gas-phase chemistry in two types of interstellar cloud (see Table 1.3b). First, we shall discuss the initiating chemistry of the most important elements in a diffuse interstellar cloud, *i.e.*, one that is easily penetrated by the mean interstellar radiation field (described in Section 3.2.4) and also by energetic cosmic rays (Sections 1.1.3 and 3.2.6). Then we shall discuss the chemistry in dark molecular clouds from which starlight is almost entirely excluded by the extinction caused by interstellar dust, and in which we shall assume that the chemistry is driven by cosmic ray ionization alone. Finally, we shall make some remarks about general methods for forming species with even greater complexity than is achieved in these chemical networks. Ref. 1 illustrates how a plausible chemical network may generate important insights into the physical nature of an astronomical source. A discussion of the initiating gas-phase chemistries in interstellar clouds and many related topics in this chapter may also be found in ref. 2, while a more elementary introduction can be found in ref. 3. The roles of dust grains in interstellar chemistry, operating side by side with gas-phase chemistry, will be described in Chapters 7 and 8.

These initiating chemical networks have emerged after decades of debate in which proposed mechanisms have been tested by laboratory measurements and theoretical studies of key rate coefficients, together with exploration of the predictions from computational models. Astrochemistry continues to involve close cooperation between astronomical observers, chemical modellers, laboratory chemists, and theoreticians working on fundamental processes. Many thousands of rate coefficients have been measured, computed, or estimated (and are continually re-assessed) in the creation of these astrochemical networks. We shall discuss the databases and libraries of reactions that support these chemistries in Chapter 5 and some model calculations made to exploit them.

4.2　The Initiating Chemistries of the Important Elements in Diffuse and Dark Clouds

4.2.1　Forms of Hydrogen in Diffuse and Dark Clouds

Since hydrogen is by far the most abundant element in interstellar space, it will drive the chemistry of carbon, oxygen, nitrogen, and other elements wherever it is able to do so, through reactions involving the neutral hydrogen atoms and molecules, and also especially in reactions with hydrogen ions. As we saw in Section 3.2.4, both hydrogen atoms and molecules are unaffected by the interstellar radiation field and consequently are largely neutral. However, cosmic rays collisionally ionize both species:

$$H + c.r. \rightarrow H^+ + e + c.r.$$

$$H_2 + c.r. \rightarrow H_2^+ + e + c.r.$$

(with minor channels $H + H$, $H + H^+ + e$, and $H^+ + H^-$)

where this second reaction is followed rapidly by

$$H_2^+ + H_2 \rightarrow H_3^+ + H$$

Although the ions H^+ and H_3^+ are minor components of the gas in both diffuse and dark regions, they play important roles in molecule formation as we'll see in the following sub-sections.

We explore the idea in this section that simple hydrides will be formed and then (in Section 4.3) that reactions with these hydrides offer reliable entry routes into the chemistry of other species.

4.2.2 Initiating Chemistry for Oxygen – Making Oxygen Hydrides

4.2.2.1 Diffuse Clouds

Diffuse clouds are observed to contain hydrogen in both atomic and molecular forms, although the relative abundances of H and H_2 may vary quite widely. Hence, the relative importance of the ions H^+ and H_3^+ in diffuse clouds may also vary. Oxygen is present in diffuse clouds as neutral atoms which react only slowly with H or H_2 at the temperatures of diffuse clouds. Oxygen chemistry occurs more rapidly through reactions with the hydrogen ions H^+ and H_3^+.

O atoms and H atoms have very similar ionization potentials (see Table 3.1), with that of oxygen being very slightly larger, the difference being an energy equivalent to 232 K. Thus, charge exchange may occur (or not):

$$O + H^+ \leftrightarrow O^+ + H$$

depending on the temperature. For low enough temperatures (compared to this critical temperature difference of 232 K), there is not enough energy released when H^+ takes an electron from the O-atom and recombines to H, ionizing the O atom, so this reaction cannot proceed in the forward direction, and any O^+ is driven back to the neutral atom. At higher temperatures, there is enough energy for the reaction to proceed in the forward direction, so O^+ is formed and reacts quickly with H_2

$$O^+ + H_2 \rightarrow OH^+ + H$$

to form OH^+.

The alternative and temperature-independent route to this product occurs in reactions of neutral O atoms with H_3^+ ions:

$$O + H_3^+ \rightarrow OH^+ + H_2$$

Successive H-abstraction reactions of OH^+ with H_2

$$OH^+ + H_2 \rightarrow H_2O^+ + H$$

$$H_2O^+ + H_2 \rightarrow H_3O^+ + H$$

followed by dissociative recombination reactions, for which the main channels are

$$H_2O^+ + e \rightarrow OH + H, \text{ or } O + 2H, \text{ or } O + H_2$$

$$H_3O^+ + e \rightarrow H_2O + H, \text{ or } OH + 2H$$

lead to the formation of the neutral hydrides OH and H_2O. The precise chemical pathway producing these two neutral molecules in diffuse clouds depends on the $H:H_2$ ratio and on the temperature structure of the diffuse cloud.

4.2.2.2 Dark Clouds

In dark clouds, most of the hydrogen is molecular, so cosmic ray ionization leads mainly to H_3^+ rather than H^+ ions. The formation of OH and H_2O in dark clouds occurs mainly by the temperature-insensitive route involving H_3^+.

4.2.3 Initiating Chemistry for Carbon – Making Carbon Hydrides

4.2.3.1 Diffuse Clouds

The ionization potential of carbon (11.26 eV) is less than that of atomic hydrogen (13.60 eV) so – as discussed in Section 3.2.4 – carbon is photoionized in diffuse clouds. However, the expected ion–molecule reaction

$$C^+ + H_2 \rightarrow CH^+ + H$$

is endothermic by about 0.4 eV (equivalent to 4640 K) and so does not proceed in cold diffuse clouds. However, this reaction does proceed in shock-heated gas, and is the source of CH^+ in these hot regions.

The initial step for carbon chemistry in diffuse clouds is a relatively slow radiative association

$$C^+ + H_2 \rightarrow CH_2^+ + h\nu$$

and this product ion reacts quickly with H_2, abstracting a H-atom:

$$CH_2^+ + H_2 \rightarrow CH_3^+ + H$$

Dissociative recombination of the important ion, CH_3^+, with electrons generates CH and CH_2:

$$CH_3^+ + e \rightarrow CH + H_2 \text{ (or 2H)}$$

$$\rightarrow CH_2 + H$$

There is also a significant dissociation channel into $C + H + H_2$ and a minor channel into CH_3. There are also other routes that produce the neutral CH_3. As well as the atomic ion, C^+, all the carbon hydrides CH_3^+, CH, CH_2 and CH_3 are important in generating further chemical complexity.

4.2.3.2 Dark Clouds

In dark clouds, the abundance of C^+ ions is very low and free carbon exists mainly in its neutral atomic form. It reacts quickly with H_3^+ to form CH^+:

$$C + H_3^+ \rightarrow CH^+ + H_2$$

and CH^+ is rapidly converted to CH_2^+ by H abstraction:

$$CH^+ + H_2 \rightarrow CH_2^+ + H$$

Thereafter, the chemistry for carbon hydrides is similar to that in diffuse clouds.

4.2.4 Initiating Chemistry for Nitrogen – Making Nitrogen Hydrides

4.2.4.1 Diffuse Clouds

Nitrogen atoms (ionization potential: 14.53 eV) are not photoionized by the interstellar radiation field in diffuse clouds, so atomic nitrogen is the dominant form in both diffuse and dark clouds. However, the reaction of atomic nitrogen with H_3^+ (that we have seen is so successful in stimulating the initial chemistry of oxygen and carbon) is endothermic and does not proceed in either diffuse or dark clouds, so nitrogen hydrides do not arise from that source. However, alternative routes to nitrogen hydrides are available. Nitrogen atoms can be collisionally ionized by cosmic rays and by He^+ ions and the reaction of N^+ ions produced in these ways with H_2 molecules

$$N^+ + H_2 \rightarrow NH^+ + H$$

provides an entry into gas phase nitrogen chemistry in diffuse clouds. This reaction is impeded by a small energy barrier (equivalent to ~40 K) that hinders the reaction from occurring at the low temperatures (~10 K) of dark clouds. However, in diffuse clouds the reaction proceeds and the NH^+ may undergo several H-abstraction reactions with H_2 molecules to form ions up to NH_4^+:

$$NH^+ \rightarrow NH_2^+ \rightarrow NH_3^+ \rightarrow NH_4^+$$

and the sequence of H-abstractions stops at that point. Dissociative recombinations and photodissociations may then generate the neutral hydrides NH, NH_2 and NH_3.

4.2.4.2　Dark Clouds

Depending on the dark cloud temperature, the formation of nitrogen hydrides in dark clouds may depend to some extent on the same sequence of reactions described above for the generation of these molecules in diffuse clouds. As we shall see below, molecular nitrogen and other N-bearing molecules may be formed by routes involving other species. Nitrogen hydrides in dark clouds are unlikely to form in exchange reactions of N atoms with other hydrides such as OH and CH, since the favoured channels lead to NO and CN. However, a route to N_2 in dark clouds exists (as we'll see in Section 4.3) and the reaction of molecular nitrogen with H_3^+ can proceed as follows:

$$N_2 + H_3^+ \rightarrow N_2H^+ + H$$

with dissociative recombination of the product ion:

$$N_2H^+ + e \rightarrow NH^+ + N$$

These reactions give a minor channel to NH^+ and thence to nitrogen hydrides, as occurs in diffuse clouds.

4.2.5　Initiating the Chemistry for Sulfur – Making Sulfur Hydrides

4.2.5.1　Diffuse Clouds

With an ionization potential of 10.36 eV, sulfur atoms are mainly ionized in diffuse clouds. However, the reaction of S^+ with H_2 to make SH^+ does not proceed at the low temperatures of diffuse clouds. A slow radiative association of S^+ with H_2 can, however, occur to form H_2S^+. A rapid radiative recombination of this ion

$$H_2S^+ + e \rightarrow SH + H$$

$$\rightarrow S + 2H$$

occurs in equal channels, with a minor channel of radiative recombination of H_2S^+ with electrons giving small amounts of neutral H_2S. Thus, the major form of sulfur in diffuse clouds, S^+, leads to small amounts of SH and very small amounts of H_2S.

However, if some neutral sulfur atoms are present as well as S^+ ions, then the temperature-independent reactions

$$S + H_3^+ \rightarrow SH^+ + H_2$$

$$\rightarrow H_2S^+ + H$$

followed by dissociative recombination reactions of H_2S^+, generate SH. The ion H_2S^+ does not abstract a H-atom from H_2 (unlike the ion H_2O^+ in the case of oxygen chemistry), so H_3S^+ cannot be formed and H_2S cannot arise from dissociative recombination of the H_3S^+ ion. Therefore, in diffuse clouds, SH is the main sulfur hydride that can arise from the two channels initiated by S^+ and S reactions.

4.2.5.2 Dark Clouds

Here, nearly all the sulfur is present as neutral atoms, and their reaction with H_3^+ ions leads to H_2S^+ ions, and thence *via* dissociative recombination to SH neutral molecules, just as in the diffuse cloud case.

4.3 Making New Molecules from Hydrides

Once simple hydrides are available, reactions of other atoms and ions with hydrides open chemical routes to a wide variety of simple molecules. One route producing carbon monoxide in diffuse clouds begins with the reaction of C^+ with OH:

$$C^+ + OH \rightarrow CO^+ + H$$

followed by

$$CO^+ + H_2 \rightarrow HCO^+ + H$$

The ion HCO^+ is a detected interstellar species, and its main loss route is in dissociative recombination:

$$HCO^+ + e \rightarrow CO + H$$

providing the important molecule CO, the second most abundant molecule in interstellar space, after H_2.

A simple barrier-free neutral exchange reaction involving OH leads to the formation of molecular oxygen:

$$OH + O \rightarrow O_2 + H$$

Carbon atoms, ions, and hydrides and their ions are important in generating a variety of species in both diffuse and dark clouds. For example, barrier-free neutral exchanges of O atoms with CH or CH_2 are important routes to CO:

$$O + (CH \text{ or } CH_2) \rightarrow CO + (H \text{ or } 2H)$$

in addition to the reaction of C^+ with OH, while the reaction of O atoms with CH_3 radicals generates formaldehyde molecules:

$$O + CH_3 \rightarrow H_2CO + H$$

At low temperatures, reactions of nitrogen atoms with the hydrides OH, CH and CH_2 can produce NO, CN, HCN, and HNC:

$$N + OH \rightarrow NO + H$$

$$N + CH \rightarrow CN + H$$

$$N + CH_2 \rightarrow HCN \text{ or } HNC + H$$

while nitrogen hydrides react with O and C atoms to form NO and CN. The formation of NO and CN may be followed by

$$N + NO \rightarrow N_2 + O$$

$$N + CN \rightarrow N_2 + C$$

providing the molecular nitrogen that enables NH^+ to be formed in dark clouds (see Section 4.2.4).

Sulfur monohydride reacts with O, C, and N atoms to produce SO, CS, and NS, but the reactions of SH with OH, CH, and NH do not proceed. Atomic sulfur may provide more effective routes to the products sulfur monoxide, carbon monosulfide, and nitrogen sulfide:

$$S + OH \rightarrow SO + H$$

$$S + CH \rightarrow CS + H$$

$$S + NH \rightarrow NS + H$$

A reaction of SO with OH generates sulfur dioxide:

$$SO + OH \rightarrow SO_2 + H$$

while neutral exchange reactions of S with CH_2 and CH_3 generate thioformyl, HCS, and thioformaldehyde, H_2CS:

$$S + CH_2 \rightarrow HCS + H$$

$$S + CH_3 \rightarrow H_2CS + H$$

4.4   Steps to Greater Complexity

In Section 4.3 we have shown that – starting with hydrogen and the next four most abundant elements in the form of interstellar atoms and (where appropriate) their ions and their hydrides – simple chemical steps of gas phase reactions allowed under cool or cold interstellar conditions produce a wide range of rather small molecules. Nearly all of the reactions and their rate coefficients and reaction products invoked in the investigation described in Section 4.3 have been studied experimentally and, sometimes, theoretically. These are viable schemes; all the molecules predicted by the descriptions in Section 4.3 are observed to exist in interstellar clouds. The important question of whether such schemes can produce not only the appropriate molecules but also the molecules in their observed abundances is one that must be tested by numerical work using computational models. It may be that although we have proposed routes to form certain species, as outlined in the preceding sections, the abundances predicted by numerical studies may be too small when compared with observational results. We shall discuss computational models of interstellar chemical networks in Chapter 5.

The schemes discussed in Section 4.3 produce molecules containing up to four atoms. Obviously, the discussion has been illustrative and incomplete, and a much wider range of predicted molecules of this size range would emerge from a more complete discussion. However, the question addressed in this section is this: how can we extend these schemes to produce molecules of even greater complexity? The Appendix to Chapter 1 shows that detected molecules range in size up to at least about a dozen atoms. The fact that very many of the detected molecules are organic molecules suggests that it is the special properties of carbon that enable greater complexity to be achieved in reaction networks of this kind.

We consider here *carbon insertion reactions*, involving either C atoms or C^+ ions. These reactions promote the growth of molecular complexity. For example, insertion of the C^+ ion into the methyl radical

$$C^+ + CH_3 \rightarrow C_2H_2^+ + H$$

forms the acetylene ion, which on dissociative recombination gives the ethynyl radical C_2H, while the insertion of the neutral C-atom into acetylene

$$C + C_2H_2 \rightarrow c\text{-}C_3H + H \text{ and } l\text{-}C_3H + H$$

generates both the linear and cyclic forms of the propynylidyne radical (both *l*- and *c*- forms are detected in the interstellar medium; *c*-C_3H is the smallest detected cyclic molecular structure in interstellar clouds). Carbon insertion into hydrocarbons generally involves the loss of hydrogen (as in these examples). Therefore, growth by these routes tends to lead to unsaturated species. Growth may also involve insertion of radicals such as C_2H, rather by atoms or atomic ions. For example, the reaction

$$C_2H + C_2H_2 \rightarrow C_4H_2 + H$$

generates a linear carbon molecule, diacetylene, C_4H_2 (or H–C≡C–C≡C–H, a *carbon chain* molecule), and reactions of molecules like diacetylene with CN generate the cyanopolyyne series $HC_{2n+1}N$ (n = 1,2,3, ...). These molecules have strong dipole moments; they radiate more readily and can be detected more easily than the pure hydrocarbons. The cyanopolyynes are found in some cool stellar envelopes and in dark clouds (up to $n = 5$).

For larger species, growth may also occur by radiative association or by so-called *condensation reactions* of two molecules into a larger entity such as

$$CH_2^+ + C_3H_2 \rightarrow C_5H_3^+ + H$$

Reactions of CH_3^+ are also very effective in producing molecules of greater complexity. These ions can be inserted into other molecular structures by radiative associations, followed by dissociative recombinations. For example,

$$CH_3^+ + H_2O \rightarrow CH_3OH_2^+ + h\nu$$

followed by dissociative recombination

$$CH_3OH_2^+ + e \rightarrow CH_3OH + H$$

and this may be an important route providing interstellar methanol. Reactions of methanol with CH_3^+ may take this complexity to another level, but the product species depends on which end of the methanol molecule is attacked by the CH_3^+ ion. One possibility is an attack on the hydroxyl end:

$$CH_3OH + CH_3^+ \rightarrow CH_3OCH_4^+ + h\nu$$

and dissociative recombination of this ion gives several channels:

$$CH_3OCH_4^+ + e \rightarrow O + CH_3 + CH_4$$

$$\rightarrow CH_3 + CH_3OH$$

$$\rightarrow CH_3OCH_3 + H$$

where the last (fairly minor) channel gives dimethyl ether. Alternatively, an attack by CH_3^+ on the methyl radical end of the molecule

$$CH_3^+ + CH_3OH \rightarrow C_2H_5OH_2^+$$

gives the ion $C_2H_5OH_2^+$ which dissociatively recombines to give the likely products (although the rate coefficient and branching ratios are not well-determined)

$$C_2H_5OH_2^+ + e \rightarrow H + H_2O + C_2H_4$$

$$\rightarrow H + H_2 + CH_3CHO$$

$$\rightarrow H + CH_3CH_2OH$$

The branching products are ethylene, acetaldehyde, and ethanol. Reactions of the types described above may obviously generate even larger species.

It is clear that there are many potential routes to molecular complexity in interstellar clouds, and it is also apparent that a relatively simple astrochemical gas-phase network can (potentially) generate a wealth of chemistry to compare with observational data. One consequence is that the computational models (on which tests of the ideas in this and earlier chapters must be made) are very extensive and require data for many thousands of reactions. We now go on to discuss in Chapter 5 how these models, their databases, and their libraries of reactions are organised. Then we shall be ready to consider if the gas-phase chemistry outlined here is able to account for the presence and abundances of all molecules detected in interstellar clouds. We shall also consider the formation of dust in circumstellar regions (Chapter 6) and the role of dust in astrochemistry (Chapters 7 and 8).

References

1. J. H. Black and A. Dalgarno, *Astrophys. J. Suppl.*, 1973, **34**, 405.
2. A. G. G. M. Tielens, *Molecular Astrophysics*, Cambridge University Press, 2021.
3. D. A. Williams and T. W. Hartquist, *The Cosmic-Chemical Bond*, RSC Publishing, 2013.

5   Databases and Computer Software for Astrochemistry in Interstellar Clouds

"The objective of the present work is to examine the principal chemical processes by which molecules are formed and destroyed by using as comprehensive a chemical reaction network as possible, and to allow for ready change and updating of the chemical reaction network as the laboratory and observational data bases improve." (S. S. Prasad and W. T. Huntress Jr.)[1]

5.1   The Need for Databases

Astrochemical models developed over more than fifty years are used to calculate the abundances of hundreds of species involved in thousands of chemical reactions. This requires the systematic collection of an enormous number of data, *i.e.*, a database. Typically, astrochemical databases contain mainly two-body gas-phase processes, although increasingly three-body reactions are also being included. Some models consider both gas-phase and grain-surface chemistry. Consequently, databases have been extended to include kinetics data for heterogeneous processes; however, we shall not discuss heterogeneous processes in this chapter (but see Chapters 7 and 8); our discussion in this chapter is restricted to gas-phase chemistry.

Astrochemistry: Chemistry in Interstellar and Circumstellar Space
By David A. Williams and Cesare Cecchi-Pestellini
© David A. Williams and Cesare Cecchi-Pestellini 2023
Published by the Royal Society of Chemistry, www.rsc.org

While the most widely used databases have broad application, others are designed for specific conditions of particular astronomical regions, such as very cold pre-stellar cores, or atmospheres of exoplanets. Some chemical descriptions are coupled to either hydrodynamics or magnetohydrodynamics.[2] Chemistry may modify thermal or transport properties, and in turn modify the hydrodynamic properties of a region, for example through viscosity or cooling rates. To be useful, networks may need to incorporate thousands, if not tens of thousands, of reaction rates, at variable temperature and pressure. In this chapter we shall focus mainly on two databases that have general, rather than specific, applications, the UMIST Database for Astrochemistry (UDfA) and the Kinetic Database for Astrochemistry (KIDA). Our discussion will assume that interstellar clouds are physically static rather than dynamically evolving.

As we have seen in Chapter 1, astrochemistry in the interstellar clouds occurs under very dilute conditions with number densities ranging from 10 to 10^{10} molecules (mostly H_2) per cm^3; gas temperatures are normally low, ~ 10 K. Most reactions involve hydrogen in some form, which constitutes more than 90% (by number of atoms) of the total number density. Hydrogen and helium account for nearly all the nuclear matter in today's Universe. Of the remaining 0.1%, about one half of the atoms are oxygen, and a quarter of them are carbon. All the other elements comprise the remaining 0.025%. Major isotopes are ^1H, ^{16}O, ^{12}C, ^{14}N, ^{20}Ne, ^{24}Mg, ^{28}Si, ^{32}S, ^{36}Ar, and ^{56}Fe. Relevant minor isotopes are deuterium (^2H), ^{18}O, and ^{13}C. Before the formation of heavy elements in the post-recombination era of the early Universe (as we shall discuss in Section 5.7), the critical chemical processes involved only hydrogen, deuterium, helium and lithium.

The chemistry under interstellar conditions does not take place in thermodynamic equilibrium, and the chemical processes are dominated by the kinetics and branching ratios of the accessible reactions (see Chapter 3). Three-body reactions are frequently unimportant, unless very high densities are attained. Because of the very low temperatures, the molecular processes, not being in thermodynamic equilibrium, require energy input to initiate. As a consequence, barrierless reactions involving ions and radicals are prevalent in the interstellar medium. As we have seen in Chapter 3, these species are readily created from the photo-ionization and photo-dissociation of more stable species by short-wavelength radiation, or by ubiquitous cosmic rays that form the primary source of energy that initiates chemical reaction networks. As nuclear spin statistics plays an important role at very low temperatures, isotopic-exchange reactions may yield various nuclear modifications in some molecular species. A highly non-statistical isotopic distribution of H_2 is in fact present in cold regions, where the polar species H_2D^+ and even HD_2^+ are observed (see Section 3.2.10). The derived abundance ratio of ND_3 to NH_3 is near $1:1000$ compared with a cosmic deuterium to hydrogen ratio of $1:100000$.

In general, ions are long-lived in low-density interstellar environments, so their chemistry plays an important role in space (see Chapters 3 and 4). Cations are most common, although anions are also present (see Section 3.2.8). Depending on the gas densities, they are generally either atomic or protonated species. Some radical cations may also be very stable, especially for resonantly stabilized species such as the family of PAHs. The protonated hydrogen molecule, H_3^+ is the most abundant molecular ion in the Universe. It plays a pivotal role in interstellar chemistry (see Chapter 4), through proton transfer reactions with neutral constituents (*e.g.*, carbon monoxide or water). These kinds of exoergic ion–molecule reactions generally proceed at collision frequency. However, this is not always the case, as in the reaction of He^+ with H_2 which is strongly inhibited, essentially not occurring at low collision energies.

As we shall see in this chapter, gas-phase reaction networks have met with great success in understanding the origin and abundances of many species detected in interstellar clouds (see the Appendix to Chapter 1 for a list of the detected interstellar and circumstellar species). However, it is also clear that these models have not always been successful, and problems remain, especially for some larger organic species. Laboratory experiments modelling both surface chemistry and the response of ices deposited on grain surfaces to starlight and to cosmic rays show that surface chemistry is responsible for simple hydride formation (including H_2) and that the chemical processing of ices may be able to account for the formation of some larger species detected in interstellar clouds. We shall discuss these topics in Chapters 7 and 8.

The perceived richness of interstellar chemistry has been growing steadily in the variety of objects and regions observed in the Milky Way and other galaxies. Some of these regions are quiescent, *i.e.*, essentially motionless, so any chemistry occurring within them has ample time to reach steady state. This takes place on relatively long timescales of the order of ten thousand to one million years. In some other cases, as for example, in the very early stages of gravitational collapse of a gas cloud to form a star, the chemistry in the gas is unlikely to attain steady state. In that case, the observed prevalent species are not necessarily those that arise in steady state.

5.2   Chemical Kinetics

Interstellar chemistry is kinetically rather than thermodynamically controlled, as is readily understood on the base of simple arguments.[3] At temperatures of a few tens of K, equilibrium calculations are straightforward. In the hydrogenation of carbon monoxide, the second most abundant molecule, to methane and water, $CO + 3H_2 \rightarrow CH_4 + H_2O$, the ratio between products and reactants $n_{CH_4} n_{H_2O} / n_{CO} n_{H_2}$ predicted at equilibrium differs from interstellar observations by more than 500 orders of magnitude. We may note that the hydrogenation step to methane formation contains a sequence of hydrogenation reactions, the first of them, $CO + H_2 \rightarrow H_2CO$, being 7 kJ mol^{-1}, *i.e.*, endothermic in the gas phase (1 kJ mol^{-1} is equivalent to 120 K). Thus, efficient conversion of CO to CH$_4$ is clearly inhibited for temperatures around 100 K or lower. In regions where high densities and temperatures are intrinsic to their evolution, such as Earth's atmosphere or the atmospheres of some stars, detailed kinetic schemes are frequently not required and simpler equilibrium prescriptions may be adequate.

5.2.1   Rate Equations

The rate of a chemical reaction is expressed as a change in concentration of some species with time. Therefore, the dimensions of the rate must be those of concentration (*e.g.*, moles per litre or m^{-3}) per unit time. The reaction A → B implies the disappearance of A or the appearance of B, so the corresponding rates are expressed as $-dn_A/dt$ and $+dn_B/dt$, respectively. The mathematical equation relating concentrations and time is called the *rate equation*. Such an equation gives the number of reactants in a single step (also known as *molecularity*). Thus, elementary reactions are referred to as unimolecular, bimolecular, or termolecular (and so on), depending on whether one, two or three species are involved in the process.

For the general reaction $\alpha A + \beta B + \gamma C + ... \rightarrow$ *products*, the rate equation is

$$\text{Rate} = k n_A^a n_B^b n_C^c \cdots \tag{5.1}$$

In this equation, k is called the *rate constant* (or rate coefficient), and a, b, and c are the *order* of the reaction with respect to the corresponding reactant. The orders are not generally coincident with the stoichiometric coefficients, except for reactions that take place in a single step. In very complex reactions the overall kinetic order loses its meaning, since the reaction rates are not simple functions of concentrations. Nevertheless, in most elementary reactions the dependence on concentration is first-order or second-order. In some reactions that involve one reactant, the rate may be independent of the concentration of the reactant (zero-order) over a wide range of values

$$-\frac{dn_A}{dt} = k(n_A)^0 = k. \tag{5.2}$$

Integration of this equation between the limits of $n_A^{(0)}$ at zero time and $n_A(t)$ at some later time, t, gives $n_A(t) = n_A^{(0)} - kt$. Zero-order reactions have been defined as chemical kinetics curiosities,[4] and are always an artefact of the conditions under which the reaction is carried out. In the first-order reaction A → B, the rate equation reads as

$$-\frac{dn_A}{n_A} = kdt \tag{5.3}$$

which together with the normalization condition $n_A(t) + n_B(t) = n_A^{(0)}$ gives the exponential decay solutions $n_A(t) = n_A^{(0)} e^{-kt}$ and $n_B(t) = n_A^{(0)}\left(1 - e^{-kt}\right)$. A reaction is second-order if the rate is proportional to the square of the concentration of one reactant (A + A → B), or to the products of the concentrations of two reactants (A + B → C). In the first case we get

$$-\frac{dn_A}{dt} = k(n_A)^2 \tag{5.4}$$

which gives the solution $n_A / n_A^{(0)} = 1/(n_A^{(0)}kt + 1)$. In the second one, the rate equation reads as

$$-\frac{dn_A}{dt} = -\frac{dn_B}{dt} = kn_A n_B \tag{5.5}$$

and the solution is

$$kt = (n_A^{(0)} - n_B^{(0)})^{-1} \times \ln\left(\frac{n_A / n_A^{(0)}}{n_B / n_B^{(0)}}\right) \tag{5.6}$$

Rate laws for reactions of higher orders may be occasionally important, such as, *e.g.*,

$$-\frac{dn_A}{(n_A)^m} = kdt \tag{5.7}$$

where m is the order of the reaction. The solution is given by the relation

$$n_A^{1-m} = (n_A^{(0)})^{1-m} + (m-1)kt \tag{5.8}$$

Practically, no elementary reactions involving more than three molecules are known, because of the very low probability of near-simultaneous collision of more than three molecules. Even termolecular reactions (three-body) appear to have a low chance of occurring, when we compare the collision time with the time between collisions. Under very favourable conditions (as in Earth's atmosphere, for example), the time between collisions is about 10^{-9} s, while collisions last 10^{-13}–10^{-12} s. Their ratio is thus approximately 0.001, which means that the probability of three molecules colliding simultaneously is of the order of $(0.001)^2$, about 1 collision in a million. It is much more likely that the process occurs as a series of bimolecular collisions.

Other types of systems may be quite common, for example, a single reactant converted into several different products simultaneously (parallel first order reactions). In this case, the rate of disappearance of A is the sum of the rates for all the processes, and the change in concentration of A with time is the same as in eqn (5.3), with $k = \sum_j k_j$, where k_j is the rate of the j-th process.

Another interesting case is that of reversible reactions A $\rightleftarrows$ B. For first order reactions we get

$$-\frac{dn_A}{dt} = \frac{dn_B}{dt} = k_f n_A - k_r n_B \tag{5.9}$$

where k_f and k_r are the rates of the direct (forward) and reverse reactions, respectively. At each time t, the sum of the transient concentrations is equal to the sum of the initial concentrations of both reactants, $n_A(t) + n_B(t) = n_A^{(0)} + n_B^{(0)}$, and eqn (5.9) reads as

$$\frac{dn_B}{dt} = k_f\left[n_A^{(0)} + n_B^{(0)} - n_B(t)\right] - (k_f + k_r)n_B(t) = -k_s n_B(t) + c \tag{5.10}$$

where $k_s = k_f + k_r$ and $c = k_f\left(n_A^{(0)} + n_B^{(0)}\right)$. When $n_B^{(0)} = 0$, the solution is

$$n_B(t) = \frac{k_f}{k_s} n_A^{(0)}\left(1 - e^{-k_s t}\right) \tag{5.11}$$

As the ratio k_f/k_r becomes progressively large, k_f approaches k_s, that is, the reaction rate becomes essentially identical to that of an irreversible first order process, see eqn (5.3). In Figure 5.1, we present the solution (5.11) in the dimensionless form $\dfrac{n_B(\tau)}{n_A^{(0)}} = \dfrac{1}{1+\delta}\left[1 - e^{-(1+\delta)\tau}\right]$ with $k_r = \delta k_f$ and $\tau = t/k_f^{-1}$, the time measured in units of the formation time, as the destruction rate (*i.e.*, δ) decreases to zero.

In general, an astrochemical model describes the evolution of the chemical composition of a mixture of gas and solids (in the form of tiny dust grains) under astrophysical conditions, possibly including

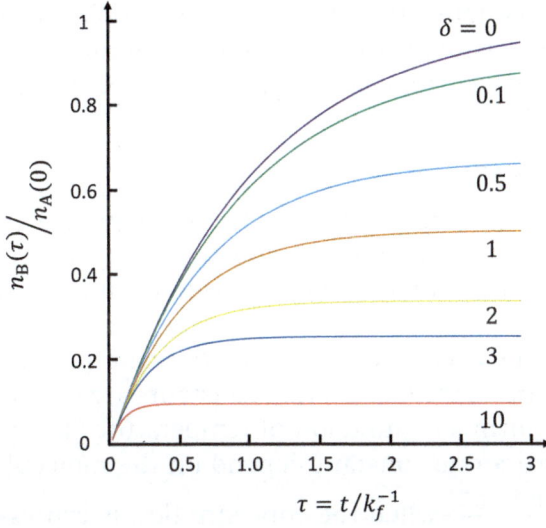

Figure 5.1 Solution for the evolution of abundances in the reversible reaction A $\rightleftarrows$ B as given in eqn (5.11). The figure is given in dimensionless form for various values of the forward and reverse reaction rates where k_f and k_r are the rates of the forward and reverse reactions, respectively, with $k_r = \delta k_f$ and $\tau = t/k_f^{-1}$.

considerations of geometry and physical dynamics. Technically, from the numerical point of view, chemical models are described by a system of differential equations, such as

$$\frac{\mathrm{d}n_i}{\mathrm{d}t} = P_i - n_i L_i \qquad (5.12)$$

with $i = 1, 2, ..., n_T$, where n_T is the total number of chemical species involved. The production (P) and destruction (nL) terms refer to all chemical and physical processes that produce and destroy the i-th species. They are therefore functions of all the n_T species included in the network of chemical reactions. Specifically, these quantities are a superposition of first-, second-, and third-order rate equations describing unimolecular, bimolecular, or termolecular elementary reactions, respectively. The time derivative should be understood as comoving (or total), *i.e.*, $\mathrm{d}/\mathrm{d}t = \partial/\partial t + (\vec{v} \cdot \vec{\nabla})$, v being the velocity of fluid, and the density-based solver as coupled to the flow and energy equations. In practice, modellers compute the evolution of the densities of species for a set of parameters, and from an initial composition (see Section 5.7). The list of parameters includes, among others, temperature, density, cosmic-ray ionization rate, elemental abundances, and the primary kinetic parameters, the reaction rate coefficients.

For very large systems, the number of equations and processes can lead to hypersensitivity to some model parameters (called order parameters) and even to bi-stability. We shall not discuss these topics.

5.2.2 Rate Constants

The rate constant defines the intrinsic velocity of a chemical reaction, and it is specific for a particular reaction at a given temperature. Rate constants are usually measured by following the concentration of species over time. The temperature dependence of k is often measured and can be used to calculate the relevant quantities characterizing a reaction. Consequently, rate constants are commonly given in a generalized form as a function of temperature. The physical units associated with a rate constant depend on the molecularity α of the reaction, $k = \dfrac{\text{rate}}{n^{\alpha}} \equiv \dfrac{n^{1-\alpha}}{t}$ when the concentration is expressed in terms, *e.g.*, of the number density in cm^{-3} (as in most databases), the units are s^{-1} for first-order ($\alpha = 1$), $cm^3 \, s^{-1}$ for second-order ($\alpha = 2$), and $cm^6 \, s^{-1}$ for third-order ($\alpha = 3$) reactions, respectively.

Two general approaches are available to correlate the rate constants with temperature. These are the *transition state theory* summarized by the Eyring equation, and the empirical *Arrhenius rate law*. Let's consider the reaction $A + B \rightleftarrows (AB)^* \rightarrow C$, with the corresponding rate equation

$$\frac{dn_C}{dt} = kn_A n_B = k^* n_{(AB)^*} \tag{5.13}$$

where $(AB)^*$ is the activated complex, also known as the *transition state*, and k^* is the rate constant of its transformation into products (see Figure 5.2), this latter term being independent of the reaction.

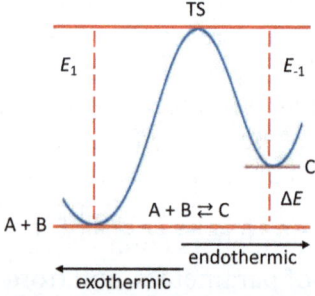

Figure 5.2 Energy transfers in the reactions $A + B \rightleftarrows (AB)^* \rightarrow C$. The transition state is at the energy maximum, and if this energy barrier can be overcome, then the reaction can proceed.

To understand qualitatively how such a chemical reaction may take place, it is assumed that a sort of (pseudo) chemical equilibrium occurs between reactants and the activated transition state complex (AB)*, through an equilibrium constant

$$K^* = \frac{k_f}{k_r} = \frac{n_{(AB)^*}}{n_A n_B} \tag{5.14}$$

where k_f and k_r are the rate constants of the direct and reverse channels in A + B $\rightleftarrows$ (AB)*. From Figure 5.2, it is clear that there is an *activation energy*, a certain amount of energy required for the reaction to occur. The transition state is formed at the top of the energy "hill". Once the energy barrier is overcome, the reaction is able to proceed and product formation occurs. The rate of the reaction (AB)* $\rightarrow$ C, here called k^*, is thus equal to the number of activated complexes decomposing to form products, that is, the frequency with which the active complex surmounts the barrier, $k^* = \kappa_T(k_B T/h)$, where k_B is Boltzmann's constant, h is Planck's constant, and κ_T is a *transmission coefficient* for overcoming a potential barrier. κ_T is often taken to be unity. Since $k = k^* K^*$, using the thermodynamic relationships

$$\Delta G^* = -RT \ln K^* = \Delta H^* - T\Delta S^* \tag{5.15}$$

we obtain the Eyring equation

$$k = \kappa_T \frac{k_B T}{h} e^{-\left(\frac{\Delta S^*}{R} + \frac{\Delta H^*}{RT}\right)} \tag{5.16}$$

In eqn (5.15) and (5.16), R is the gas constant, and ΔG^*, ΔH^*, and ΔS^* are the Gibbs energy, enthalpy, and entropy of activation, respectively. Relation (5.16) correlates the rate constant with temperature, and thereby allows the experimental determination of the activation parameter ΔG^* from the temperature dependence of the reaction rate.

The Arrhenius rate law was derived empirically before the development of the transition state theory, from the observation that the rates of reactions increased exponentially as the absolute temperature increases

$$k = A e^{-\frac{E_a}{RT}} \tag{5.17}$$

with A being the pre-exponential factor, and E_a the Arrhenius activation energy (see Figure 5.2). The Arrhenius equation defines a macroscopic rate coefficient, and therefore it avoids any mechanistic considerations, including whether one or more intermediates are involved in the overall conversion of the reagents into the products. On the other hand, the Eyring equation works at a microscopic level, depending on a value of ΔG^* at any single step of the conversion of a reactant to a product. For elementary reactions, the Eyring and Arrhenius descriptions are simply connected through the relations $E_a = \Delta H^* + RT$, and $\Delta S^* = 4.576(\ln A - 10.753 - \ln T)$. It is important to note that E_a computed from the Arrhenius equation is not a thermodynamic quantity, and differs from ΔH^*, a thermodynamic quantity, by the term RT. When data are obtained over a sufficiently wide temperature range, the Arrhenius equation often does not describe them accurately, and the plots of $\ln k$ $versus$ $1/T$ show a deviation from a straight line. A better approximation is given by

$$k = AT^{\beta} e^{-\frac{E_a}{RT}} \tag{5.18}$$

where the three parameters A, β, and E_a can be varied to fit the data over a wide range of temperatures. Arrhenius-type relations, see (5.18), are employed in most databases to describe the rates for standard bimolecular gas-phase reactions.

It is a major endeavour to derive reaction rates and incorporate them into chemical reaction networks. In fact, the number of chemical processes in astrochemical models for which rate constants are known at low interstellar temperatures is extremely small. Generally, rate coefficients are measured at room temperature. Their extrapolations down to interstellar conditions have often given misleading results. Moreover, when reaction reactants may be converted into several different products simultaneously, the branching ratios for these different channels have frequently not been determined, even though the rate constant of the global reaction is known. Unfortunately, in many important cases, neither experimental nor theoretical work has been performed, and the values of rate constants exploited in models are uncertain, made up of a combination of experiment, theory, and speculations founded on intuition and common sense. Sometimes such an approach may be the only way to estimate a rate coefficient, but it may lead to wrong conclusions.

Additional problems are posed by certain types of species, $e.g.$, radicals, metastable isomers, and ions, that would never emerge on our planet. A molecule can survive for many years before it collides with another molecule in interstellar space, so even if it's not very stable, it can exist for a long time (weeks or months). On Earth, it would almost immediately react with something else to transform into a more stable species. In such circumstances, experiments are difficult to perform, and the only viable alternative is a theoretical approach, exploiting a suite of techniques grouped together under the term *computational chemistry*. Understanding chemical bonding and reactivity has been the principal goal of theoretical chemistry since the advent of quantum mechanics, more than one century ago. The theoretical estimate of a rate coefficient proceeds through the calculation of the so-called *potential energy surface*, an energetic "landscape" in which different directions represent the geometrical parameters of molecules in the process of transforming themselves into products, and the height of the "land", the energy associated with these geometrical parameters. A fundamental requirement for the reaction to proceed is that there is an electronically and thermodynamically allowed (or at least not strongly forbidden) path that leads from the reactants to products. Typically, the energetic landscape of a chemical reaction shows two probability peaks, appearing as minima of the potential energy surface, that correspond to the reactant and the product states (see Figure 3.1). Figure 5.3 shows the potential energy surface of a fictional system as a function of two quantities, R_1 and R_2, that may be regarded as *space coordinates*; these quantities generally describe a few selected degrees of freedom, and are often referred to as collective variables.

The minima corresponding to reactants or products may be connected by paths (see the red line), along which rearrangements and

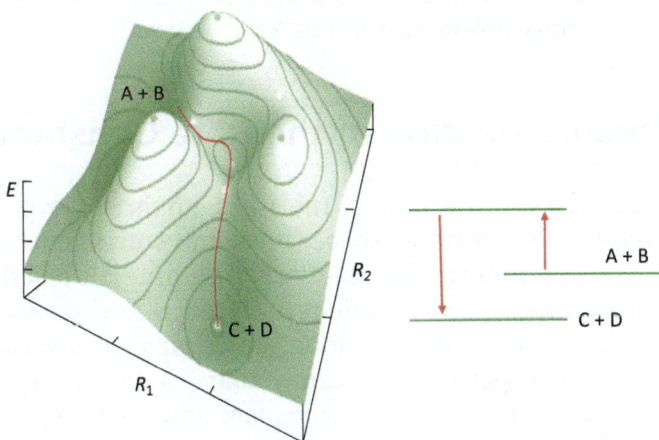

Figure 5.3 Potential energy surface of a fictional system. The figure shows schematically the potential energy surface of a fictional system as a function of two quantities, R_1 and R_2, that may be regarded as *space coordinates*; these quantities generally describe a few selected degrees of freedom, and are often referred to as collective variables.

reactions can occur. The maximum along the path is the transition state, which as we have seen above corresponds to the minimum energy required to transition between two minima on the potential energy surface, and plays a crucial role in the description of chemical transformations. Transition states are first order saddle points, a maximum in one coordinate and minima in all others. The longitudinal section of the potential energy surface along the maximum of the transition state is shown in Figure 5.2. The identification of such energetically favourable reactive paths is achieved through an iterative exploration of the potential energy surface, applying a perturbation represented by a sum of Gaussians centred along the trajectory followed by reactants, thus forcing the system to migrate from one minimum to the next. This approach is the core of a powerful technique called *Metadynamics*,[5] which has been designed to overcome the serious problem of remaining stuck in only one probability maximum (typically corresponding to an energy minimum) for the duration of the run. While such state-of-the-art computational methods give a realistic and accurate description of chemical reactivity, they have the drawback of having very high computational costs, requiring enormous amounts of computer time, memory, and storage space. Because chemical models contain thousands of chemical reactions, countless rates are still very poorly defined. Model predictions can be consequently strongly affected by uncertainties in the rate coefficients. Reducing the degree of uncertainty intrinsic in databases is therefore a primary problem (Section 5.6).

5.3   Parametrization of the Rate Constants

We have described in Chapter 3 the most important types of gas-phase reactions occurring in space, involving some general classes of reactions: (1) unimolecular reactions including all reactions of a species with photons and cosmic-rays, leading to dissociation and ionization; (2) bimolecular reactions, describing all chemical reactions between two species; (3) termolecular reactions, requiring the collision of three particles at the same place and time, *i.e.*, third order reactions. Before discussing the current and most important databases, we describe how rate constants are parametrized. Usually, in a database, rate coefficients are stored in the form of a three-parameter set, conventionally written as $\alpha, \beta,$ and γ (not to be confused with other uses of these symbols). This set can be used to compute the rate coefficients using different formulas depending on the process considered, and the range of temperatures desired. Although basic descriptions are similar, different databases may have their own prescriptions in mapping non-standard reaction rate constants, such as those involving ion-polar systems. While the vast majority of reactions can be parametrized in this way, some critical reactions, *e.g.*, H_2 formation on grain surfaces or the photodissociation of H_2 or CO (see Section 3.2.5), would be oversimplified in nonspecific parametrizations, and they may be described by means of *ad hoc* recipes.

5.3.1 Unimolecular Reactions

Photodissociation and photoionization are unimolecular reactions; their rates in s^{-1} are given by the relation

$$k = 4\pi \int_{v_{th}}^{v_M} F_v \sigma(v) dv \tag{5.19}$$

where v is the frequency, v_{th} the threshold frequency of the process, and v_M the maximum frequency of the adopted radiation field. In eqn (5.19), F_v is the intensity of the radiation field in photons per cm^2 s^1 Hz^1 sr^1, and $\sigma(v)$ is the frequency-dependent cross section of the photoprocess in cm^2. F_v depends upon the boundary intensity $F_v^{(0)}$, *i.e.*, the unattenuated background field, and the solution of the radiation transfer problem, which in its simplest form (the Beer–Lambert law) may be written as $F_v = F_v^{(0)} e^{-\tau_v}$, with τ_v being the optical depth in the integration frequency range. In the interstellar medium, the relevant spectral range is the vacuum ultraviolet band, with $v_M = v_L$, where v_L is the frequency of the Lyman limit of atomic hydrogen, corresponding to a photon energy of 13.6 eV (see Section 3.2.4). Dust is the major source of opacity (see Section 1.1.4). The optical depth may be written in terms of the extinction as $\tau_v = A_V/1.086$ (see Section 1.1.4). The rate constant therefore turns out to be

$$k = 4\pi F_v^{(0)} \int_{v_t}^{v_L} e^{-A_V/1.086} \sigma(v) dv \approx \alpha \exp(-\gamma A_V) \tag{5.20}$$

where γ is a parameter included to take into account the increased extinction in the ultraviolet compared to the extinction at visible wavelengths, A_V. Finally, the parameter α is the photorate for $A_V = 0$, *i.e.*, at the region boundary. A better, not frequently used parametrization is given by

$$k = \alpha \exp(-\beta A_V - \gamma A^2_V) \tag{5.21}$$

In astronomical regions with complex geometry, the simple expression (5.20), or (5.21), may not be applicable, and eqn (5.19) needs to be solved accordingly. In such a case, cross-sections rather than rate constants are necessary (see Section 5.5).

In the class of unimolecular reactions, there are also two types of cosmic-ray interaction with the gas. The first is direct cosmic-ray ionization with a rate given by

$$k = 4\pi\eta(1+\varphi)\int j(E)\sigma(E)\mathrm{d}E \tag{5.22}$$

where $j(E)\mathrm{d}E$ is the cosmic-ray proton differential spectrum, $\sigma(E)$ the cross-section for ionization by a fast charged particle of energy E, and η and φ correction factors to include the contributions to ionization of heavy nuclei in cosmic rays and secondary electrons, respectively. Assuming that cosmic-rays lose their energies only marginally in penetrating a cloud of gas and dust, the ionization rate constant becomes

$$k = a\zeta \tag{5.23}$$

with ζ being the H_2 cosmic-ray ionization rate, frequently used as a reference, and a a constant depending on the chemical species.

Interacting with the gas, cosmic-rays may produce excitations rather than ionizations. When the target species is molecular hydrogen, cosmic rays may excite efficiently the upper states of the Lyman and Werner systems of this molecule. Both these systems are located around 10 eV above the ground state ($B^1\Sigma_u^+$ and $C^1\Pi_u$, see Figure 3.6). The emissions resulting from these excitations may maintain a significant flux of chemically effective ultraviolet photons in the interior of dense clouds, where interstellar radiation may not reach (see Section 3.2.6). Such cosmic-ray induced photo-processes proceed at a rate

$$k = a\zeta\left(\frac{T}{300\,\mathrm{K}}\right)^\beta \frac{1}{1-\omega} \tag{5.24}$$

where a is the probability per cosmic-ray ionization that the photoreaction takes place, and ω the average dust grain albedo, which measures the efficiency by which dust particles scatter radiation in the ultraviolet (see Section 1.1.4).

5.3.2 Bimolecular Reactions

As we have noted earlier (see eqn (5.18)), two-body reactions are parametrized by the Arrhenius type expression (also known as the Arrhenius–Kooij formula)

$$k = \alpha \left(\frac{T}{300\,\text{K}} \right)^{\beta} \exp\left(-\frac{\gamma}{T} \right) \tag{5.25}$$

in which α is the pre-exponential factor, β the temperature dependence of the correction factor, and γ the activation energy or energy barrier of the reaction in units of degrees K. Bimolecular reactions describe several gas-phase processes: ion-neutral, charge exchange, mutual ion neutralization, dissociative recombination, radiative recombination, associative electron detachment, neutral–neutral, and radiative association. In the rest of this chapter, the concentrations of the species involved in the chemistry are in the form of number densities measured in cm^{-3}. Therefore, the rate constant in eqn (5.25) has the physical units $\text{cm}^3\,\text{s}^{-1}$.

5.3.3 Termolecular Reactions

A three-body reaction is a gas-phase reaction of the form

$$A + B + M \rightleftarrows AB + M$$

M is an unspecified collision partner that carries away excess energy to stabilize the molecule (forward direction) or supplies energy to break the bond (reverse direction). Three-body processes become important at high densities, never below $10^{10}\,\text{cm}^{-3}$ (see Section 3.2.11). The rate law of these reactions is given by

$$rate = k n_A n_B n_M \tag{5.26}$$

with k given in $\text{cm}^6\,\text{s}^{-1}$. Different species may be more or less effective in acting as a collision partner, as summarized by a collision efficiency ε for each species, so the concentration of the third reaction partner is $n_M = \sum_j \varepsilon_j n_j$. Since collision efficiencies may be incorporated into the rate constants, n_M is frequently taken as the total number density. Under certain conditions, the rate of reactions described by the expression (5.26) may depend on pressure as well as temperature. When reaction rates behave as first-order in M at low pressure, but become zero-order in M as concentration increases, such pressure-sensitive reactions are called *fall-off reactions*.

5.3.4 Interaction of Gas Phase Species with Dust Grains

The interaction between gas-phase species and dust grains is described by desorption and accretion processes. This latter event is also called *freeze-out* to indicate the removal of species from the gas-phase, and their attachment onto the ultracold surfaces of dust grains. In the following we derive the rates for accretion and thermal desorption. Gas-phase species may either physisorb on the grain surface by van der Waals forces, or form a chemical bond with the lattice in a process called chemisorption. Occasionally, accreting species might interact with other adsorbed species, giving rise to weakly bound van der Waals complexes (*e.g.*, $HCN\cdots CO$), more strongly bound hydrogen-bonded clusters (*e.g.*, $(H_2O)_2$ and $H_2SO_4\cdots H_2O$), radical complexes (*e.g.*, $OH\cdots O_2$), and ionic clusters (*e.g.*, $H_3O^+\cdots(H_2O)_n$). The rate of accretion of a gas-phase species ($cm^{-3}\ s^{-1}$) is determined by the collisional frequency of the species with a grain, multiplied by a sticking efficiency S

$$k^{(a)}n_g = S\langle \pi a^2 n_d\rangle v_g n_g \qquad (5.27)$$

where n_g is the number density of a gas phase species, a the radius of a dust grain with number density n_d, and $v_g = \sqrt{8k_B T / \pi m_g}$ the average gas-phase thermal velocity, with m_g being the mass of the colliding gas-phase species. The term $\langle \pi a^2 n_d\rangle$ represents the sum total cross-sectional area of *all* grains of all radii per unit volume of space, assuming the grains are spherical. The sticking efficiency is the probability of a species to adhere to the surface upon a collision, and it depends on how well the incoming species can dissipate its kinetic energy.

The desorption term represents the ability of a species to depart from the grain surface, ending up back in the gas phase. Various desorption processes are possible, and usually a distinction is made between thermal and non-thermal processes. Species on or near the surface of a dust grain can be desorbed by photons of the interstellar radiation field, if the grains are not too heavily shielded by surrounding dusty gas. In dark regions, desorption may be induced directly by the passage of heavy cosmic rays, or indirectly by the weak radiation field created when cosmic rays ionize hydrogen atoms and molecules. A multitude of other mechanisms is possible, including reactive or chemical desorption, where the excess heat generated upon reaction allows desorption of the products.

The thermal residence time of a molecule or atom on the dust grain surface is predominantly determined by its thermal desorption rate:

$$k^{(d)} = v_0 \exp(-E_b/k_B T_d) \qquad (5.28)$$

where E_b is the binding energy of a species to the surface, T_d the dust temperature, and ν_0 the characteristic vibration frequency for each adsorbed species. According to eqn (5.28), the rate constant has physical units s^{-1}, and describes a unimolecular reaction. Binding energy and vibration frequency are two important parameters in astrochemical models. The following estimate is frequently adopted for the characteristic frequency:

$$\nu_0 = \sqrt{\frac{2N_s E_b}{\pi m}} \qquad (5.29)$$

where N_s is the number density of surface sites per dust grain, and m the mass of the desorbing species. Thermal desorption rates are frequently assumed to be first-order reactions, as in eqn (5.28). However, this is only strictly valid in the sub-monolayer regime, *i.e.*, when the species is bound on top of the ice surface. In practice, the situation is much more complicated, and reactions of different order may occur. Zero-order desorption, *i.e.*, a constant desorption rate, generally takes place when multiple layers of the same species are deposited, while second-order desorption, *i.e.*, a quadratic dependence of the rate on the number of surface species, occurs when desorption proceeds through chemical ejection of species that are formed *via* a second-order surface reaction.

5.3.5 Isotopic Exchange Reactions

As mentioned briefly in Section 3.2.10, isotopic exchange reactions are ion–molecule reactions that can transfer isotopes between molecules. At the low temperatures prevailing in interstellar regions, many quantum effects such as zero-point energy difference between reactants and products, large rotational level spacings and nuclear spin effects become important. The rate constants of this class of reactions are given in terms of eqn (5.25). However, while these reactions are exothermic in the forward direction, they become endothermic by the same amount in the reverse direction. This leads therefore to strong fractionation effects at low temperatures. At the typical gas temperatures of molecular clouds, $T \approx 10$ K, the reaction $H_3^+ + HD \rightarrow H_2D^+ + H_2$ becomes important, giving rise, upon reactions with abundant neutral species (*e.g.*, CO and N_2), to deuterated molecular ions (*e.g.*, DCO^+ and N_2D^+). The reaction of H_3^+ with HD proceeds further along the exothermic sequence $H_2D^+ \rightarrow D_2H^+ \rightarrow D_3^+$. All these species provide important information on the physical conditions in dense and cold interstellar regions. An isotopologue differs from its parent chemical in that at least one atom has a different number of neutrons, thus changing the reduced nuclear mass in a molecule, affecting vibrational and rotational level energies, and in some cases modifying the nuclear spin (which may imply the loss of symmetry properties).

5.4 Databases

Making explicit the production and destruction terms in eqn (5.12), the rate law of a single molecular species in the gas-phase reads as

$$\frac{dn_i}{dt} = \sum_{j,l} k_{jl} n_j n_l + \sum_l k_l n_l - n_i \left(\sum_{l \neq i} k_{il} n_l + k_i \right) + k_i^{(d)} n_i^s - k_i^{(a)} n_i \tag{5.30}$$

where n_i and n_i^s are the gas-phase and solid-state concentrations of the i-th species, k_i and k_{ij} the rate constants for first-order and second-order kinetics, and $k_i^{(d)}$ and $k_i^{(a)}$ the desorption and accretion rates, assumed to describe unimolecular reactions.

Chemical models consist of a large number of rate laws such as the one described by eqn (5.30), resulting in a set of coupled, non-linear ordinary differential equations, to be integrated forward in time to generate molecular abundances at different epochs. To be useful, such a numerical model must include a collection of species and an extensive reaction scheme connecting them that adequately describes the complexity of the different regions of the interstellar medium.

In the following, we shall discuss some well-known databases and the ways in which these databases are organized. As the most popular types of databases, they consist of rows and columns in a series of tables, although KIDA provides a user-friendly interface optimized for specific requirements.

5.4.1 The UMIST Database for Astrochemistry (UDfA)[6]

UDfA, the UMIST Database for Astrochemistry, originally developed at the University of Manchester Institute of Science and Technology (UMIST), was launched in 1991. It was the world's first public chemical kinetic database for astrochemistry. The latest release, the fifth named *RATE12*, was made in 2012.[7] The gas-phase chemistry is described by 6173 reactions among 467 species – including 268 cations, 28 anions and 171 neutral species – composed of 13 elements. Unimolecular processes are identified by having a photon (photoionization and photodissociation, "PHOT"), a cosmic-ray proton (direct cosmic-ray ionisation, "CP"), and a fluorescent photon (cosmic-ray induced photo-processes, "CRP") as the second reactant, respectively.

Table 5.1 List of all reaction types reported in UDfA and KIDA.

Type	Description	UDfA	KIDA
Direct cosmic ray process	Dissociation/ionization by cosmic-ray impacts	CP	CR
Photo-process induced by cosmic-rays	Photo-dissociation and ionization by UV photons emitted by H_2	CR	CRP
Photo-process	Dissociation/ionization by absorption of radiation (UV/X-rays)	PH	Phot
Bimolecular reaction	Neutral–neutral	NN	Bimo
	Ion–neutral	IN	Bimo
	Anion–cation (mutual neutralization)	MN	Bimo
	Dissociative neutral attachment	NN	Bimo
Charge exchange reaction	$A^+ + B \rightarrow A + B^+$	CE	CE
	$A^+ + B^- \rightarrow A + B$	MN	
Radiative association	Association stabilized by the emission of a photon	RA	RA
Associative detachment	$A^- + B \rightarrow AB + e^-$	AD	AD
Electronic recombination and attachment	Dissociative neutral recombination	DR	ER
	Attachment of an electron	REA	ER
	Emission of a photon	RR	ER
Collisional dissociation	$A + BC \rightarrow A + B + C$	CD	Bimo
Termolecular reaction	Third-body assisted	CL	3-Body

Photons are also involved as product (emission) in bimolecular radiative processes, radiative association (RA), radiative electron attachment (REA), and radiative recombination (RR). The list of all reaction types is reported in Table 5.1.

The database is contained in a file, in the form of colon-separated item columns:

reaction no.:type:R1:R2:P1:P2:P3:P4:NE:α:β:γ:T_l:T_u:ST:ACC:REF

where "type" in the second field indicates the type of reaction, R1 and R2 are reactants, P1–P4 products, and α, β, and γ, the fit parameters. When a rate constant has a complex dependence on temperature, a single set of parameters may not be enough, and the fields and columns 10–17 are repeated the appropriate number of times, given in the 9th field "NE", for each temperature range between T_l and T_u, the 13-th and 14-th fields, respectively. Finally, for each set of parameters, the field "ST" provides information about the way a rate constant has been derived, the final field "REF" references the source of the data, either papers or DOIs,[†] and "ACC" is a code representing the error. As an example, let's take reaction no. 707 in *RATE12*,

$$S + C^+ \rightarrow C + S^+$$

[†]A DOI, or Digital Object Identifier, is a string of numbers, letters and symbols used to permanently identify an article or document and link to it on the web.

This is a "type = CE" (*i.e.*, charge transfer, see Table 5.1) reaction, and its rate constant is known in the "NE = 2" temperature ranges, the first being $T_1 - T_u = 10-499$ K, and the second $500-50\,000$ K. Since this reaction is bimolecular, the rate is parametrized through eqn (5.25)

$$k_1 = 5 \times 10^{-11} \text{ and } k_2 = 5.54 \times 10^{-12} \left(\frac{T}{300\text{K}}\right)^{0.86} \exp\left(-\frac{680.7\text{K}}{T}\right)$$

in the first and second temperature ranges, respectively. Both rates have been measured ("ST = M"), while their accuracy is "ACC = highly uncertain" (which means that the rate is just an estimate) and "ACC = <25%", respectively.

The database contains some additional kinetics data, to be exploited for specific tasks, such as the accurate modelling of deuterium chemistry in very cold sources. This can increase the number of reactions by about an order of magnitude, as well as the number of differential equations to be solved. The four molecular ions H_3^+, H_2D^+, D_2H^+, and D_3^+ have *ortho* and *para* forms corresponding to the spin states of the protons (in H_3^+, H_2D^+) or the deuterons, D^+ (in D_2H^+, D_3^+). Deuterons, unlike protons, have spin $I = 1$ and hence obey Bose–Einstein statistical laws; as a consequence, D_3^+ has one additional nuclear spin symmetry, *i.e.*, *meta* ($I = 1,3$ and $g_I = 10$), together with the usual *ortho* ($I = 1,2$ and $g_I = 16$), and *para* ($I = 0$, $g_I = 1$) states. We need to consider all the symmetry forms of a molecule as different, separate species, because the exo/endothermicities of the forward and reverse reactions involved in deuteration depend on the symmetry of the reactants and products.

Another separate file contains a few tens of termolecular reactions, listed using the same format as *RATE12*, but including a third column for reactants. The rate constants are parametrized using the Arrhenius–Kooij formula. Since these processes are termolecular reactions, rate constants should be multiplied by the density of the third body (frequently the total number gas density) to ensure that the rate is in the right units.

5.4.2   The Kinetic Database for Astrochemistry (KIDA)[8]

KIDA is a database of chemical reactions used to study the chemistry in different types of astrophysical media, such as the interstellar medium, protostellar envelopes, protoplanetary discs, and planetary atmospheres. According to the intentions of the authors, this database differs from the UDfA or other available sources[‡] in that many

---

[‡]*E.g.*, the OSU database initially developed by Prasad and Huntress (1980),[1] updated by Chun Leung and Eric Herbst,[9] and then revised by Herbst and collaborators in the subsequent years.[10]

reactions have an associated *data sheet*, in which are presented the reasons for the choice of particular values of the rate coefficients and their uncertainties. KIDA is an evolving database as new data and corrections are constantly added, checked and certified. Although users can collect information and rates for specific reactions in order to create their own collections of chemical reactions, there are some selections of gas-phase chemical reactions extracted from KIDA that can be directly used in astrochemical models. Most of these chemical reaction sets have been developed for dense, cold regions, and include also deuterium fractionation. The file *kida.uva.2014* contains 6992 chemical reactions for a total of 7506 rate coefficients, among 489 different species, composed of 13 elements. The included reactions are: photo-processes, bimolecular reactions, dissociative neutral attachments, charge exchange reactions, radiative associations, associative detachments, electronic (dissociative) recombinations and attachments, direct cosmic-ray processes, photo-processes induced by cosmic-rays, and third-body assisted associations. As in UDfA, there are ancillary species, *i.e.*, Photon, CR, and CRP representing photons, cosmic-ray particles, and ultraviolet photons induced by cosmic-ray H_2 excitations. These species always appear as "reactants" in unimolecular reactions. Photons may also act as "products" in radiative associations. A list of all reaction types used in KIDA is reported in Table 5.1, together with the UDfA analogues.

Further networks have been added by various research groups. They include gas–grain chemical networks, descriptions of surface chemistry, with some including the Eley–Rideal mechanism (see Chapter 7), the formation of complexes at the surface of the grains, gas–grain interactions, photon-dominated region chemical networks, and bimolecular and trimolecular reactions in planetary atmospheres. There is, also, a version of the gas-phase OSU network (see footnote [‡]) for temperatures up to 800 K and a list of reactions involving the generic species PAH, considered as a small grain, in two charge states, anion and neutral.

Reactions between ions and polar neutral species are more complex than those involving non-polar neutrals, because the attraction between a charge and a rotating permanent dipole is no longer a central force motion. An example is the radiative association HCN + $CH_3^+ \rightarrow CH_3CNH^+ + h\nu$. The rate coefficients for ion-polar systems are described by two non-standard expressions, valid in the low and high temperature regimes. These temperature ranges are defined through the dimensionless parameter $x = \mu_D / \sqrt{2\alpha k_B T}$, where μ_D is the dipole moment in units of Debye, and α_P the scalar polarizability in units of

10^{-24} cm^3; $x = 2$ marks the boundary between the two regimes. The rate coefficient for an ion–dipole system is given in terms of the Langevin rate, $k_L = 2.34 \times 10^{-9} q \sqrt{\alpha_p / \mu}$, in which q is the electronic charge, and μ the reduced mass of the colliders. The rate constant becomes

$$k = \alpha\beta \left[0.62 + 0.4767\gamma \left(\frac{300\,\text{K}}{T} \right)^{1/2} \right] \quad (5.31)$$

when $x \geq 2$, and

$$k = \alpha\beta \left[1 + 0.0967\gamma \left(\frac{300\,\text{K}}{T} \right)^{1/2} + \frac{\gamma^2}{10.526} \frac{300\,\text{K}}{T} \right] \quad (5.32)$$

otherwise. In these equations, α represents the branching ratio of the reaction, β is the Langevin rate, and γ represents the value of x at 300 K. Being bimolecular reactions, the rate coefficients have the physical units cm^3 s^{-1}.

In the KIDA rate files, reactions are organized in rows: 2 reactants, 4 products, 3 parameter fits for the rate constants (α, β, and γ), the uncertainty factor on the rate coefficient (F), the temperature dependence of this uncertainty factor (g), the type of uncertainty, the type of reaction (itype), the range of temperatures in which the rate coefficient is valid, a number referring to the formula needed to compute the rate coefficient of the reaction, the number of the reaction in the chosen network, the number of triplets α, β, and γ, *i.e.*, the number of times the reaction is listed in the network, and, finally, the recommendation about the use of that specific rate constant. For instance, for the charge transfer reaction $S + C^+ \rightarrow C + S^+$ that we used as a test case in UDfA, we get

---

1.500×10^{-9}	$0.000e + 00$	$0.000e + 00$	$2.00e + 00$	$0.00e + 00$	logn5	10	499	3	5066	1 1
5.540×10^{-12}	8.578×10^{-1}	6.807×10^2	$1.50e + 00$	$0.00e + 00$	logn5	500	50000	3	5066	2 2

This reaction is no. 5066, it is known in two temperature ranges, $T_l - T_u = 10$–499 K and 500–50 000 K, "itype = 5" indicates an exchange reaction, and the fitting formula is the Arrhenius–Kooij formula (5.25) indicated by the number 3. The statistical distribution representing the uncertainty is Lognormal (logn), which means that 68% of the results fall within one standard deviation and 95% fall within two standard deviations; the rate is known up to a multiplicative factor $F = 2$ and 1.5 in each temperature range, while the uncertainty does not depend on the temperature, $g = 0$. In the first temperature range there is no recommendation (value = 1), while in the second one the value has been validated by experts over the temperature range (value = 2). For this specific reaction, $S + C^+ \rightarrow C + S^+$, the UDfA and KIDA databases use similar data, although the accuracy estimate is less severe in the KIDA database. This suggests a certain degree of arbitrariness in the evaluation process, namely "the assignment of these uncertainties is a subjective assessment of the evaluators".

As we noted above, in KIDA it is possible to export groups of reactions involving elements or species, either single or multiple. As an example, including all options in the dropdown menu, the search for the species HCN returns two isomers, HCN and HNC, and 10 unimolecular, 329 bimolecular, 5 termolecular, and 9 surface reactions. Data can be handled removing unwanted reaction partners, elements, reaction types, and so on, and eventually exported in the format described above, after having set the temperature range.

5.4.3 Ancillary Data. The NIST Database

For each species included in a reaction database, additional information is frequently included. This may encompass important molecular characteristics, such as common and stoichiometric formulae, mass, polarizability, surface binding energy, dipole moment, and enthalpy of formation.

Two important sources of data are Alexander Burcat's Thermodynamic Data[11] and the NIST (National Institute of Standard and Technology) database.[12] The "Burcat" database provides thermochemical data in the field of combustion and related topics, for about 2500 species, mainly organic molecules and radicals, but also inorganic species connected to combustion and air pollution. More extensive sets of data are found in the NIST database, and in particular in the NIST Chemistry WebBook[13] and Computational Chemistry Comparison and Benchmark DataBase (CCCBDB),[14] where in addition to chemical (*e.g.*, reaction thermochemistry), thermodynamic and thermophysical data, we may extract information on phase change, infrared and mass spectra, vibrational and/or electronic energy levels, X-ray photo-electron spectroscopy, and fluid properties, such as sound speed, viscosity and thermal conductivity. Many other listed, important properties are, *e.g.*, microwave spectra, electron impact ionization cross-sections, and gas-phase kinetics. An example regarding the helium hydride ion HeH^+ is reported in Table 5.2. For many species of

Table 5.2 List of main physical and chemical properties of HeH^+.

Name	Helium hydride cation	KIDA
Mass (amu)	5.00988	NIST Chemistry WebBook
Polarizability (10^{-24} cm^3)	0.15	NIST CCCBDB
Dipole moment (Debye)	−1.304	NIST CCCBDB
Enthalpy of formation (kJ mol^{-1})	1353.391 ($T = 0$ K)	Burcat
	1357.834 ($T = 298$ K)	Burcat

astrophysical relevance, a comprehensive list of molecular or atomic characteristics can be found already collected in the KIDA database, selecting the chosen species.

5.4.4 Chemical Kinetics and Photochemical Data for Use in Atmospheric Studies. The JPL Database[15]

This is a list of kinetics and photochemical data compiled by the NASA Panel for Data Evaluation, established in 1977 by the NASA Upper Atmosphere Research Program Office, with the aim of providing a critical assessment of the reliability of kinetic and photochemical data to be used in models of atmospheric chemistry. Cyclically, every two to three years, updates to this evaluation are released in electronic form. The latest evaluation, number 19, was released in May 2020.

The compilation includes bimolecular and termolecular reactions, photochemical cross-sections (discussed in the next section), data for equilibrium chemistry, heterogeneous processes including reactions on surfaces and uptake of molecules into aerosols and liquid droplets. When this last process takes place within the atmospheric aqueous phase, it is more properly considered as involving homogeneous aqueous-phase reactions, including reactions taking place in clouds, fogs, aqueous aerosols, and raindrops. The compilation also includes thermodynamic data, among which are entropy and enthalpy of formation values at 298 K for a number of atmospheric species. While rate constants for bimolecular reactions are given by eqn (5.25) with $\beta = 0$, rates for termolecular and other pressure-sensitive reactions are expressed through a combination of Arrhenius type relations.

5.5 Photochemistry and Radiation Chemistry

When the geometry of an astronomical region cannot be simply described by a slab or a sphere, as commonly done for general interstellar clouds, and/or the energy of the incoming photons exceeds the Lyman continuum, the solution to eqn (5.19) is no longer reasonably approximated by means of the expression (5.20). It is therefore necessary to integrate directly eqn (5.19). In such a case the needed data are cross-sections instead of rate constants. In addition to photochemistry describing processes initiated by photon-induced electronic excitation not involving ionization, we must consider chemical changes produced by the absorption of radiation of sufficiently high energy to produce ionization. Such processes are described (see Section 3.3) as radiation chemistry, involving all ionizing types of radiation, high-energy photons (gamma radiation and X-rays), charged particles, and neutrons. Here, we limit the discussion to X-rays. The distinction between photo- and radiation chemistry is somewhat arbitrary and made uncertain by the use of vacuum ultraviolet and multiphoton photoionization, and the use of X-rays to study excited states of hydrocarbons.

A unique feature of radiation chemistry is the production of a cascade of low-energy secondary electrons. These secondary processes are typically much more important than the corresponding ionization, excitation, and dissociation events caused directly by photons or particles. This may also lead to new reaction pathways not available to photochemistry, such as, *e.g.*, dissociation induced by electron attachment, $AB + e \rightarrow A + B^-$ or $A^- + B$. Moreover, reaction cross-sections can be several orders of magnitude larger for electrons than for photons. The interaction of X-rays with matter proceeds by raising an inner-shell electron to an empty valence shell, or ejecting it from the atom (the most likely result). In the process, an electron drops into the vacancy from a higher orbit and a second electron takes up the energy which is used to eject it from the atom. This additional emission of electrons is called the *Auger effect*, and the emitted electrons are called Auger electrons (see Figure 5.4). Higher energy X-rays are mainly absorbed by photoionization from the K and L shells of heavy atoms, resulting in the production of multiply charged ions due to the Auger effect. In a dense gas illuminated by X-rays, the chemical structure and thermal balance are completely determined by the radiation field. In particular, X-rays producing significant ionization fractions drive an efficient ion–molecule chemistry (see Section 3.2.7).

PHIDRATES[16] (Photo Ionization/Dissociation RATES) is a database providing a set of photoionization, photodissociation, and photodissociative ionization cross-sections and the associated photon energy-weighted excess energies of the various photo products. Data are available for 140 species, including neutral atoms, atomic ions, diatomics and triatomics, and larger molecules containing up to 10 atoms. When measured or computed cross-sections are not available, a reasonable approximation to the X-rays' interaction with heavy molecules can be obtained by assuming that the molecular X-ray absorption cross-section is the sum of the corresponding atomic cross-sections. A complete set of analytic fits to the nonrelativistic photoionization cross sections for the ground states of atoms and ions of elements from H through Si, and S, Ar, Ca, and Fe can be found in ref. 17.

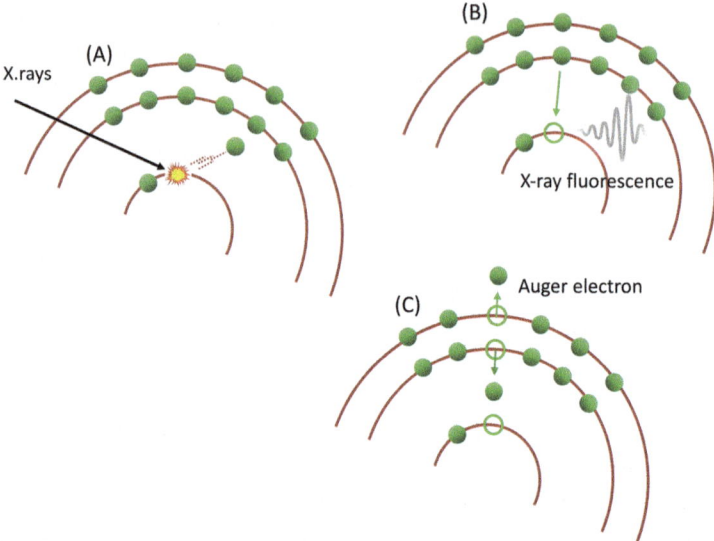

Figure 5.4 The Auger effect: the figure illustrates schematically that higher energy X-rays are mainly absorbed by photoionization from the K and L shells of heavy atoms (stage A). Sometimes an upper electron drops to fill the vacancy, followed by the emission of an X-ray photon (X-ray fluorescence, stage B), while other times such downward transition results in the production of multiply charged ions (Auger effect, stage C).

5.6   Uncertainties in the Rate Constants

The uncertainties in the rate constants of many gas-phase reactions may exceed an order of magnitude. Such structural ambiguity in chemical databases affects our attempts to reproduce the observed abundances of chemical species, blurring their dependence on the physical and evolutionary parameters of their host astronomical regions, as inferred by the solutions to the chemical network eqn (5.12).

Leaving aside exotic statistical methods involving artificial intelligence techniques, two possible aspects of the problem can be considered. First, rate constants are reasonably known within a range of uncertainty, defined through, *e.g.*, a simple standard deviation or a full probability density function; such information can be used to transfer the uncertainty from data inputs to model outputs. Second, usually only a handful of the species included in a database are considered in the interpretation of astronomical observations, so it would in principle be possible to identify key parameters for which a better estimate would reduce significantly the model uncertainties. This procedure is called sensitivity analysis.

In the KIDA database there are four kinds of statistical distributions representing the uncertainty: *normal* (or gaussian), *uniform* (with equally likely outcomes), *lognormal*, and *loguniform*. Related to the statistical distributions are two factors F and g (see Section 5.4.2), with the first describing the uncertainty factor on the rate coefficient, the meaning of which and its units depend on the adopted statistical distribution, while the second one, measured in K, is used to parametrize a possible temperature-dependence of the uncertainty. In the case of a lognormal distribution, if the rate value k_0 is known within a multiplicative factor F (no units), then the probability that the rate k is in the range $k_0/F \leq k \leq k_0 \times F$ is 68%. In other words, F is the geometric standard deviation, the exponentiated value of the standard deviation of the log-transformed values for the lognormal distribution. If we are using a loguniform distribution, then the probability reaches 100%. The first choice implies that the mean value k_0 is the preferred value. The temperature-dependence of the uncertainty factor is described by the function $F(T) = F_0 \times \exp(g|1/T - 1/T_0|)$, where $F_0 = F(T_0)$. When $g = 0$, the uncertainty factor has no (known) dependence on the temperature.

Using the distributions for F and g, all rate coefficients of a system may be randomly varied within a certain range, generating a set of evolutionary solutions for the chemical system. Since there are typically thousands of rate constants, a straightforward trial and error approach may be impractical. A simple way to overcome this problem is to implement the method through Monte Carlo sampling. Using the results of calculations, sensitivity analysis may then be performed, evaluating the correlation coefficients between inputs, the rate coefficients, and outputs of the model, the molecular concentrations.

5.7 Realizations of Gas-phase Chemistry

We have described in the previous sections the chemical processes, the management of their rates in databases, and the structure of the equations constituting interstellar, gas-phase chemical models. These aspects can be best illustrated and understood in terms of specific astronomical cases based on the methods and models presented in this chapter, for cases in which gas-phase chemistry plays a central role.

One of the cases we shall consider is the chemistry in the period between the emergence in the Universe of the first neutral atoms, about 100 000 years after the Big Bang, and the formation of the first shining objects, a few hundred million years after the Big Bang. Since there were no sources of light in the Universe, such a period is called the *cosmic dark ages*. This epoch is crucial for our understanding of cosmology as this was when the first structures formed through gravitational instability, in particular the first stars, galaxies and super-massive black holes. The early Universe was dust-free until the first generation of stars started making dust through nuclear fusion, so this cosmic period provides a perfect test case for pure gas-phase chemistry.

We have seen in Chapter 1 that, in the local Universe, the bulk of the interstellar gas is cool, relatively dense, dark (that, in this context, means mostly opaque to stellar radiation), and mainly molecular (dark and molecular clouds). Diffuse and translucent clouds are warmer but still fairly cool, less dense, rather transparent to stellar radiation, and of varying H/H_2 balance. Where H_2 is less abundant, the role of H_3^+ is consequently restricted. In that case, as we have seen in Chapter 3, cosmic rays drive the chemistry by creating H^+ ions from H atoms. Accidental resonance charge exchange between H^+ and O atoms generates O^+ ions (see Section 4.2.2) which fuel the oxygen hydride chemistry, as described earlier. Being scarcely affected by dust, the mean interstellar radiation field gives a high rate of photodissociation and photoionization, so relative molecular abundances tend to be low in such regions. In these environments, the dust–gas coupling and grain surface chemistry are negligible, so the chemistry is occurring primarily in the gas-phase.

5.7.1 Chemistry in the Cosmic Dark Ages: the First Molecules in the Universe

According to current cosmology, the Universe began about 13.8 billion years ago. In the very first moments after the Big Bang, the Universe was extremely hot, dense, and small. As the Universe expanded and cooled, conditions became just right to give rise to the building blocks of matter, quarks and electrons. These sub-atomic particles had been interacting with a background of very high energy electromagnetic radiation. In a few millionths of a second, quarks—the elementary building blocks of the heavy particles—formed protons and neutrons, which within minutes combined to form the nuclei of atoms. This matter may interact with radiation and is largely made by baryons, namely protons, neutrons and all the objects composed of them (the atomic nuclei). It is thus called *baryonic matter*. This term, in principle, should exclude particles such as electrons and neutrinos which are actually *leptons*. In astronomy, however, the term baryonic matter is used more loosely, referring to all objects made of "normal" atomic matter (the basis of everything we can see and touch), simply ignoring the presence of electrons which account for only 0.05% of the mass. Neutrinos, on the other hand, are correctly considered by astronomers to be non-baryonic. Baryons, however, do not tell the whole story, as a consistent fraction of the mass of the Universe is made of an unknown type of matter called *dark matter*, which is completely invisible. It emits no light or energy and thus cannot be detected by conventional sensors and detectors.

Once nuclei were produced, however, the temperature was still so high that atoms could not exist, and the baryonic matter was distributed as a highly ionized plasma. From that moment, the evolution of the Universe began to slow down. It took 380 000 years for most of the electrons to be trapped in orbits around nuclei, forming the first atoms. This is known as the *epoch of recombination* (even though atoms and electrons had never been combined before). The elemental composition of the gas was primarily hydrogen and helium, with traces of deuterium, ^{3}He, and lithium. Despite the absence of heavy elements, a surprising degree of chemical complexity has been proven to be possible. Until recombination, the Universe was opaque to electromagnetic radiation because photons were scattered by free electrons, being absorbed and re-emitted in a variety of unpredictable directions. As recombination occurred, the density of free electrons diminished greatly, leading to the decoupling of matter and radiation, and the Universe became transparent to light. After decoupling, the temperature of background radiation continued to decline with cosmic expansion, with the photons finally free to travel through the Universe. We can collect these photons today in the form of long wavelength, low energy radiation; they are the farthest and oldest light any telescope can detect. Such radiation, the so-called *cosmic microwave background radiation*, has an almost perfect black body spectrum with peak wavelength in the millimetre range, representing a temperature of about 2.725 K. The analysis of such relic radiation has surprisingly shown that photons and baryonic matter were not all there was in this very early Universe. The mass (and its energy equivalent) in the Universe is composed of several other components, the elusive dark matter, an even more mysterious form of energy called *dark energy*, and some lesser ingredients, including neutrinos. While photons and neutrinos make a negligible contribution, the proportions of mass/energy attributed to the various components are 68.3% dark energy, 26.8% dark matter, and 4.9% baryonic matter. Baryonic matter is therefore a relatively minor component, but it is only through observations of it in the form of stars and galaxies that the role of the obscure major constituents can be traced.

Astronomers often use the concept of red shift, z, to specify evolutionary age in the Universe. Red shift refers to the increase in wavelength of radiation received on Earth from a distant source. The increase arises because the Universe is expanding as a result of the Big Bang. Thus, an object emitting light is receding with a velocity increasing with its distance from us, as described by the Edwin Hubble's 1929 relation (also known as the Hubble law). The redshift is one of the most basic concepts of astronomy, and is one of the few observational parameters that can be measured directly. A red shift of zero means that the radiation comes from a nearby region and there is no increase in wavelength, while a positive value means the source of radiation is very distant such that the expansion velocity is great. The greater the redshift, the greater the distance the light has traveled, the younger the object that emitted the radiation we are detecting.

The recombination era began at a redshift of about $z = 2500$ when the Universe was about 100 000 years old,[§] and the temperature was about 4000 K. The interaction between the gas and the cosmic background radiation played an important role in regulating the chemistry of the gas. Helium recombination was the first to occur at a redshift of $z \sim 2000$ ($t \sim 0.13$ My), while hydrogen was still largely ionized. The first reaction to occur in the Universe was thus the radiative association forming the helium hydride ion

$$H^+ + He \rightarrow HeH^+ + h\nu \qquad (5.33)$$

So HeH$^+$ was the first molecule that ever formed in the Universe. This primacy may perhaps be disputed by He$_2^+$ which is formed by the radiative association of He with He$^+$. HeH$^+$ cannot exist on Earth, except in a laboratory where it was synthesized in 1925, and, despite its importance in models of the chemical evolution of the early Universe, for decades it went undetected in space. Eventually, using an airborne telescope, astronomers reported its detection in a planetary nebula, the final, brief stage in the life of a medium-sized star like our Sun (see Section 6.2.2). Located 3000 light-years away near the constellation Cygnus, this planetary nebula, called NGC 7027, hosts the physical conditions that allow this molecule to form. Although HeH$^+$

[§]According to the cosmological time calculator at http://www.astro.ucla.edu/%7Ewright/Cosmo-Calc.html

is of limited importance today, the chemistry of the Universe began with this ion. The previous lack of detection of this molecule called into question our understanding of chemistry in local plasmas and even our models of the early Universe. The unambiguous detection of HeH^+ in NGC 7027 provides a solid foundation for the chemical evolution of the Universe in the previous 13 billion years.

Hydrogen chemistry in the early Universe was very different from that in the local Universe, because of the absence of dust. Under such conditions, molecular hydrogen can be formed by rather slow gas-phase reactions. The formation of HeH^+, see eqn (5.33), was followed by the sequence of radiative association of H and H^+ to give H_2^+, enabling the formation of molecular hydrogen *via* charge exchange:

$$H + H_2^+ \rightarrow H^+ + H_2 \tag{5.34}$$

Although the conversion of HeH^+ to H_2^+ by reaction with hydrogen atoms may have acted as a very first source of molecular hydrogen, reaction (5.34) is a much more efficient production channel. Destruction of molecular hydrogen occurs through the reverse of reaction (5.34) and by dissociative attachment, $H_2 + e \rightarrow H^- + H$. As in the local Universe, H_2 was the most abundant molecular species during the dark ages. H_2^+ can be destroyed by photodissociation and dissociative recombination, $H_2^+ + e \rightarrow 2H$. The formation of H_2 only became effective when the radiation temperature dropped below 4000 K and photodissociation of H_2^+ ended. In this phase, the abundance of H_2 increased quite rapidly, giving rise to HD, H_3^+, and H_2D^+. The formation of HD, the second most abundant molecule, is dominated by deuteron exchange with H_2:

$$D + H_2 \rightarrow HD + H \tag{5.35}$$

and

$$D^+ + H_2 \rightarrow HD + H^+ \tag{5.36}$$

Reaction (5.36) is also the major source of HD in diffuse interstellar clouds in the local Universe. Protonation of molecular hydrogen led to H_3^+, while H_2D^+ formation proceeded through

$$HD + H_3^+ \rightarrow H_2D^+ + H \tag{5.37}$$

and

$$HD^+ + H_2 \rightarrow H_2D^+ + H \tag{5.38}$$

Major sources of HD^+ were the radiative associations of D with H^+, and D^+ with H. For all species, the main destruction channel was through energetic photons from the background radiation. The abundance of hydrogen-bearing molecules levelled off at a redshift of $z \sim 1000$ ($t \sim 0.43$ My), when the ionization fraction was still $\sim$10%. Although the ionization fraction dropped to $\sim$1% at $z \sim 800$ ($t \sim 0.63$ My), the electron concentration was still enough to provide a turnover followed by a decrease to a minimum of molecular ions through electron dissociative recombination.

As the ionization fraction froze out, the chemistry slowed down, with all molecular species forming at a much lower rate up to $z \sim 300$ ($t \sim 3$ My). At later times, $z \sim 100$ ($t \sim 16.6$ My), H^- ions, whose binding energy is 0.754 eV, were no longer effectively removed in photodetachment reactions by the photons of background radiation (the radiation temperature was below 1000 K), and promoted the formation of H_2 molecules through a two-step process initiated by the radiative attachment

$$H + e \rightarrow H^- + h\nu \tag{5.39}$$

followed by the associative detachment

$$H + H^- \rightarrow H_2 + e \tag{5.40}$$

with the increase in H_2 concentration, the abundance of most species rose and reached a peak value at $z \sim 60$ ($t \sim 36$ My).

Lithium chemistry started at $z \approx 450$ ($t \sim 1.6$ My) when singly-charged Li ions became neutral through electron recombination, $Li^+ + e \rightarrow Li + h\nu$, and mutual neutralization with hydrogen anions, $Li^+ + H^- \rightarrow Li + H$. The first Li-bearing molecule to form was LiH, mainly *via* radiative association:

$$Li + H \rightarrow LiH + h\nu \tag{5.41}$$

Due to its small binding energy (0.14 eV), the formation of LiH^+ by means of the radiative association of Li with H^+ was delayed until redshift was below $z \sim 40$ ($t \sim 66$ My). Later on, its abundance reached that of LiH.

For $z \lesssim 40$, the molecular concentrations of the most abundant species H_2 and HD "froze out" (remained unchanged), as their chemical formation times became longer than the Universe expansion time. The H^- concentration steadily dropped due to its conversion into H_2 molecules. When the redshift fell below $z = 20$ ($t \sim 180$ My), the dark ages ended, and the first structure formation in the Universe began to be important. Only H_2 and HD reached their steady state abundances, while all the other molecules remained out of equilibrium. Shown in Figure 5.5 are the computed fractional abundances (relative to the total number of baryons) of the main molecular neutrals and ions formed in the early Universe as a function of redshift; the chemical data used in this model are from ref. 19.

As is clear from the figure, most of the gas was still in atomic form, and only a tiny fraction—one hydrogen molecule out of one million hydrogen nuclei—formed the first molecules in the Universe. So why were molecules so important? The first generation of stars is thought

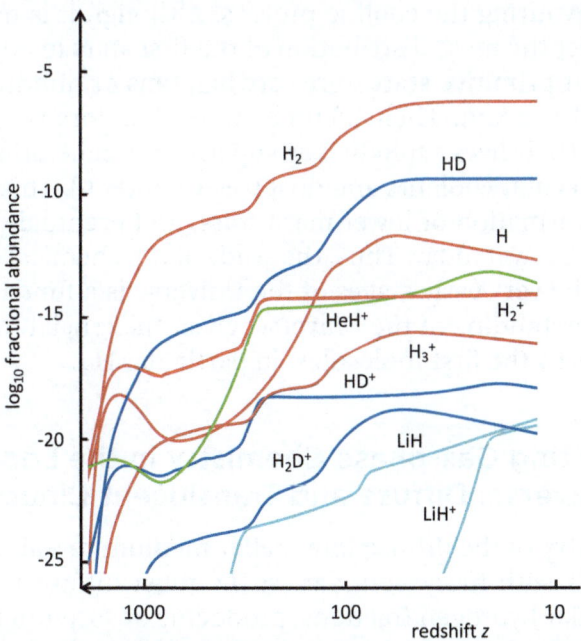

Figure 5.5 The computed fractional abundances (relative to the total number of baryons) of the main molecular neutrals and ions formed in the early Universe as a function of redshift, *z*. At redshifts of 1000, 100, and 10, the corresponding ages of the Universe were 0.434 My, 16.633 My, and 0.478 Gy.

to have emerged from baryonic matter sinking into collapsing and merging dark matter minihaloes of approximately one million solar masses. In order that stars actually formed, the atomic gas must have been able to collapse. For collapse to occur, the temperature in the collapsing region must be held low, so that the consequent pressure rise that would inhibit the collapse was suppressed. When the temperature of the gas is lower than $10\,000$ K, cooling by lines of atomic species such as the Lyman α line of the atomic hydrogen transition $H(2p) \rightarrow H(1s) + h\nu$ is ineffective. The molecules H_2, HD and LiH are much more important coolants at low temperatures due to their rotational lines generated by the transitions H_2 ($J = 2 \rightarrow 0$) at ~500 K, HD ($J = 1 \rightarrow 0$) at ~100 K, and LiH ($J = 1 \rightarrow 0$) at ~30 K. Molecules were thus able to cool down primordial clouds to a few hundred K. As the density in the condensations increased due to collapse, H_2 formation was boosted by 3-body processes:

$$H + H + H \rightarrow H_2 + H \tag{5.42}$$

further favouring the cooling process. Although it is not yet possible to predict the mass distribution of the first stars in any detail, it is thought that primitive stars were very big, tens or hundreds of times the mass of the Sun. Their lifetimes were therefore very short, and they ended their lives exploding as supernovae, generating heavy elements, molecular coolants and dust (see Chapter 6). This opened the way to the formation of lower mass stars, and eventually to the Universe as we know it today. Thus, the study of the chemical evolution of the gas in the very early stages of the Universe is a fundamental step in our understanding of the Universe, given the crucial role played in this process by the first molecules, in particular H_2.

5.7.2 Testing Gas-phase Chemistry in the Local Universe: Diffuse and Translucent Clouds

The chemistry of the diffuse interstellar medium reveals three major components with fuzzy boundaries: the most diffuse material with low molecular hydrogen fractions, producing absorption primarily by certain light hydride species (*e.g.*, CH), a slightly denser phase with a higher H_2 molecular fraction, in which heavier molecules appear (including HCN, CS, and H_2CO), and translucent clouds in which carbon is present in neutral, ionized, and molecular forms simultaneously. Many diffuse clouds in the interstellar medium, in which chemistry is dominated by rapid photoionization and photodissociation by stellar radiation, are quiescent; *i.e.*, essentially motionless, so any chemistry occurring within them has ample time to reach steady state. Temperatures are as low as 100 K or lower. These regions are perfect test cases for gas-phase chemistry.

Cold chemistry generates OH, C_2, and CH effectively, but fails to generate sufficient CO, CN, CH^+, SH^+, and HCO^+, by large factors. Further uncertainty on the nature of diffuse clouds is generated by the discovery along diffuse lines of sight of large concentrations of H_3^+, a result that is apparently inconsistent with predictions based on a high removal rate *via* electronic dissociative recombination in a cold gas. The basic problem for the chemistry is how to incorporate the ambient oxygen and carbon into molecules. Oxygen is ionized by slow charge transfers with H^+, while the abundant ionized carbon does not react rapidly with H_2, having a barrier of ~5000 K. Neutral oxygen reactions with H_2 to form OH are likewise endothermic by about 3000 K.

The need for energy sources in excess of the average energy density of diffuse and translucent clouds took into account localized volumes of warm gas created and sustained by dissipation of gas kinetic energy, as in shocks or turbulence. The classical picture of the diffuse interstellar medium has been modified, incorporating small amounts of warm gas intimately associated with the cold neutral phase. These tiny warm regions might promote the first steps in the chemistry of diffuse clouds, not only by producing the required concentrations of CH^+ through the ion–molecule exchange reaction (see Section 4.2.3)

$$C^+ + H_2 \rightarrow CH^+ + H \tag{5.43}$$

but also by inducing the formation of HCO^+, which thus leads to the formation of CO *via* dissociative electron recombination. These warm tiny regions may also provide a viable unified way to account for the observed large concentrations of H_3^+. In fact, the main H_3^+ destruction channel in diffuse clouds, dissociative electronic recombination, decreases significantly with increasing gas temperature, suggesting that transient non-equilibrium chemistry at high temperature might solve the problem. In Figure 5.6 we show the dependence of the rates (taken from the KIDA database) of the exchange reaction (5.43) and the electronic dissociative recombination of H_3^+ as functions of the gas temperature. While the first rate is totally negligible until the gas temperature reaches 2000 K, the latter decreases a factor of 20 in going from $T = 10$ to 3000 K.

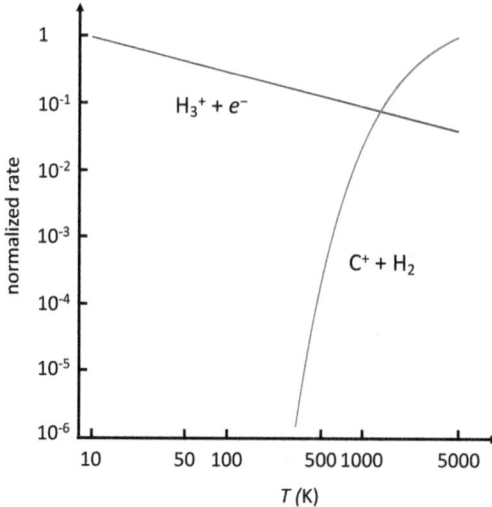

Figure 5.6 Dependence of the normalized rates (taken from the KIDA database) of the exchange reaction $C^+ + H_2$ and the electronic dissociative recombination of H_3^+, as functions of the gas temperature. While the first rate is totally negligible until the gas temperature reaches 2000 K, the latter decreases by a factor of 20 as the temperature rises from $T = 10$ to 3000 K.

In this section we present a highly idealized and extremely simple model of diffuse/translucent interstellar regions that still retains a reliable representation of the ongoing chemistry. It is often the case that a small number of factors are qualitatively and quantitatively more significant than the others, so including those factors in our description can provide an excellent starting point for building a more complete model. First, the geometry of the interstellar cloud must be defined, *i.e.*, its shape and size. Gas density and temperature must also be chosen; they can be functions of the position within the cloud, ultimately determined by heating and cooling processes, or they may be uniform. Then we need to choose the elements present in the gas and their concentrations (*e.g.*, the solar ones, as in Table 1.4), generally taken relative to hydrogen. The drivers of the chemistry, *i.e.*, the intensity of the illuminating radiation field and the level of cosmic ionization, are determined by modelling the chemistry in specific interstellar regions, and comparing the results of the model with observations. Finally, using absorption or emission spectra, we may also constrain the density and gas temperature. As dust controls the photochemistry, we may also need to take into account its albedo, and its density relative to the gas, the so-called gas-to-dust ratio, typically defined by the ratio between the total hydrogen column density N_H and the visual extinction A_V, see eqn (1.2).

The simplest type of solution is the equilibrium solution, when all the species reach their steady-state concentrations. It can be obtained by solving algebraically system (5.12), or (5.30) after having suppressed the last two terms, with all the time derivatives set equal to zero. The time-dependent solution of system (5.12) is considerably more demanding computationally, as the solution of a large number of coupled "stiff" ordinary differential equations is required. There is no unique definition of "stiff" in the literature. Generally, stiffness can be described as having time constants in a model (in our case the rate constants) that vary by several orders of magnitude. There is extensive literature on the solution of stiff ordinary differential equations, to which the reader is referred. One of the best known is Gear's algorithm.[18] Variations of this algorithm are implemented in many generally used astrochemistry codes. The solution of system (5.12) tends to approach the equilibrium solution when $t \rightarrow \infty$ (*i.e.*, if we integrate long enough). It is the equilibrium solution because it does not depend on time, so all the derivatives $dn_i/dt \sim 0$. The equilibrium solution is independent of the initial conditions, *i.e.*, it is stable.

We use an idealized description for a diffuse interstellar cloud, by considering a plane parallel slab of uniform gas density n_H and temperature T, illuminated on both sides by the ultraviolet interstellar radiation field at the location of the cloud in the Galaxy. The slab is infinite in the vertical direction, and its transverse thickness is measured in terms of the extinction produced by the dust in the visual A_V, at the cloud centre. The geometrical thickness of the "cloud" is inferred by means of the assumed gas-to-dust ratio R, through which we derive the total transverse column density as $N_H = R \times 2A_V$. Since for a cloud with constant gas number density $N_H = n_H \times L$, the cloud size becomes $L = R \times 2A_V/n_H$. We initially describe a diffuse cloud with a central extinction of $A_V = 1$, $n_H = 100$ cm^{-3}, and $T = 100$ K. Adopting the rather standard value $R = 1.9 \times 10^{21}$ cm^{-2} mag^{-1}, see eqn (1.2), we obtain $L \sim 12$ pc, where we have used the definition of parsec reported in the preliminary material at the start of this book. We consider 4 elements, hydrogen, helium, carbon, and oxygen, that (with the exception of helium) are among the main components of observable interstellar molecules. For all these elements, we use solar abundances, and we list them as fractions relative to the abundance of hydrogen: $f_H = 1$, $f_{He} = 0.098$, $f_C = 2.14 \times 10^{-4}$, and $f_O = 5.75 \times 10^{-4}$ (see Table 1.2). In this way the number density of carbon nuclei is $n_C = 2.14 \times 10^{-4} n_H \sim 2.1 \times 10^{-2}$ cm^{-3}. We consider the intensity of the radiation field as the one adopted in UDfA and KIDA, so that we can use the rates listed in those databases. We also set the cosmic-ray ionization rate to $\varsigma = 1.4 \times 10^{-17}$ s^{-1}, and the dust albedo to $\omega = 0.5$. With such choices, the chemical evolution is solved independently of the physical conditions, which are taken as fixed and homogeneous, providing thus a pseudo time-dependent evolution of the chemical concentrations.

The second step in constructing the model is the definition of the species—composed of the selected elements—whose abundances are under study, and the network of chemical reactions connecting them. In order to limit the extent of the chemical network, we restrict the number of species to $N_s = 36$, electrons included. The list is reported in Table 5.3, and it contains molecules up to 4 atoms, of which 13 are

Table 5.3 A summary of the physical and chemical parameters of the toy models in Section 5.7.2.

Geometry	Plane-parallel	Extinction (mag) at the cloud centre	1, diffuse; 3, translucent;
Dust albedo, ω	0.5	Gas-to-dust ratio, R (cm^{-2} mag^{-1})	1.9×10^{21}
Size (pc), L	12, diffuse 7, translucent	Hydrogen number density (cm^{-3}), n_H	100, diffuse; 500, translucent;
Radiation field	UDfA/KIDA	Cosmic-ray ionization rate (s^{-1}), ς	3×10^{-17}
4 elements, fractions w.r.t. hydrogen (see Table 1.2)	H 1 He 0.0977 C 2.14×10^{-4} O 5.75×10^{-4}	Kinetic temperature (K), T	100, diffuse; 30, translucent;
Number of species, N_s (electrons included)	36	Number of reactions	441

List of species included in the model			
1 atom (12)	2 atoms (13)	3 atoms (8)	4 atoms (3)
H	H_2	H_3^+	CH_3^+
He	H_2^+	CH_2	H_3O^+
C	CH	CH_2^+	$C_2H_2^+$
O	CH^+	H_2O	
H^+	OH	H_2O^+	
He^+	OH^+	C_2H	
C^+	CO	C_2H^+	
O^+	CO^+	HCO^+	
H^-	C_2		
C^-	C_2^+		
O^-	O_2		
e^-	O_2^+		
	OH^-		

neutral species, 18 are cations, and 4 are anions. The reaction network includes: 380 binary reactions, whose rates are described by eqn (5.25), (5.31) and (5.32); 37 photoreactions that given the assumed geometry are suitably approximated by eqn (5.20); 9 cosmic-ray ionizations, proceeding at rates given by eqn (5.23); 13 cosmic-ray induced photo-processes, whose rates are given by eqn (5.24). Although dust is not involved in the chemistry, we implicitly assume that the conversion of atomic hydrogen to the molecular form occurs on dust grains. The process governing the formation of H_2 is a complex one (see Section 7.3). However, it can be described quite accurately by the simple expression (s^{-1})

$$k_f(H_2) = 3 \times 10^{-17} n_H \tag{5.44}$$

The whole reaction network is composed of 440 reactions. These reactions are listed in the Zenodo repository (https://doi.org/10.5281/zenodo.6546367), where we present data from both UDfA and KIDA databases. In the calculations reported below, we exploit UDfA data. In addition to the involved species, Table 5.3 summarizes all the assumptions determining the model, geometry, boundary conditions, the elements and their abundances, the reaction types, and the initial conditions.

For these values of the parameters, we expect a scarce molecular synthesis, including molecular hydrogen. Since hydrogen is by far the most abundant element (helium is not very reactive), we can approximate the evolution of gas-phase atomic and molecular hydrogen independently of the other elements, reducing the chemistry to the simple system

$$H + H(grain) \rightarrow H_2 + grain$$
$$H_2 + h\nu \rightarrow 2H \tag{5.45}$$

where H_2 formation occurs in surface reactions (see Section 7.3), and the destruction proceeds through a two-step process involving line absorption (Section 3.2.5). System (5.45) represents a standard reversible reaction, as described in this chapter (Section 5.2.1) by eqn (5.9). The solution, eqn (5.11), reads as

$$n_2(t) = \frac{k_f n_H^2}{2k_f n_H + k_d} \left(1 - e^{-(2k_f n_H + k_d)t} \right) \tag{5.46}$$

where $n_1 = n(H)$ and $n_2 = n(H_2)$ are the number densities of atomic and molecular hydrogen, respectively, $n_H \sim n_1 + 2n_2$, and the gas is considered to be initially fully atomic, *i.e.*, $n_2(0) = 0$. k_f is given in eqn (5.44), while for k_d we use the rate given in the KIDA database, $k_d = 5.68 \times 10^{-11} \exp(-4.18A_V)$ s^{-1}. Expression (5.46) tends to the stationary solution $n_2 \sim n_H/2$ for large gas densities, $k_f n_H \gg k_d$. In the diffuse case, where the reverse is true, $k_f n_H \ll k_d$, eqn (5.46) becomes

$$n_2(t) = \frac{k_f n_H^2}{k_d}\left(1 - e^{-k_d t}\right) \tag{5.47}$$

with the stationary value, $n_2 \sim k_f n_H^2/k_d$, reached at time $t_s \geq 2/k_d$. In the present case, by substituting the values of the parameters we are using in building the model, we obtain $n_2 \sim 0.35$ cm^{-3}, and a time $t_s \sim 70\,000$ years. The gas, thus, is expected to be mainly in atomic form.

Since in the interstellar medium, the radiation field has a cut-off in energy at 13.6 eV, a suitable set of initial conditions for the integration of system (5.30) is obtained by considering all the elements with ionization potential lower than that of hydrogen in the (single) ionized form, and the remaining neutral. Thus, H, He (24.59 eV), and O (13.62 eV) are neutral, while C (11.26 eV) is ionized. The chemical evolution at the centre of our model cloud is shown in Figure 5.7.

We found, as predicted by eqn (5.47), that H$_2$ reaches its steady state concentration followed by the rest of the molecules in about 70 000 years. For longer times, chemical abundances do not vary anymore. We also show in the figure a case in which initially all the hydrogen in the gas is incorporated in H$_2$, and how the system relaxes to the same stationary state as in the previous case, in which the starting

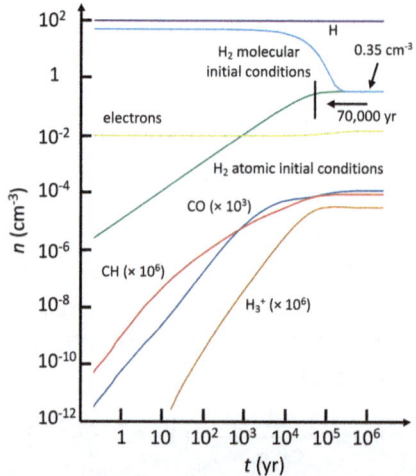

Figure 5.7   The chemical evolution at the centre of the model diffuse cloud. The chemical data used in this model are from ref. 6.

composition of the hydrogen was fully atomic. Molecular fractions, $f_X = n_X/n_H$, are very low, as evidenced by the values for H_2 and CO, the two more abundant molecular species, namely $f_2 = 3.5 \times 10^{-3}$ and $f_{CO} = 1.2 \times 10^{-9}$, both of them very far from the elemental abundances of H, O, and C (incorporated in atoms, H and O, and ions, C^+). Hydrides as CH and CH^+ remain at a level about 2–3 orders of magnitude lower than CO.

H_2 abundances depend on the gas density and the size of the cloud, and for very diffuse clouds (like the one we are considering) they are not underestimated by the toy model exploited here. On the contrary, the derived abundances of CO, CH, CH^+, HCO^+ and other species are too small (Table 5.4), showing that cold chemistry alone fails in providing the observed molecular concentrations. Additional amounts of these species may be produced during the local dissipation of turbulence that may form, in a cold cloud, pockets of hot gas. The intermittent nature of turbulence in space and time induces dynamical processes that are locally large enough to drive the heating of the gas and trigger the endothermic reactions at the base of carbon and oxygen chemistries. The lifetime of perturbation is generally as short as a few hundred to a thousand years, much smaller than the lifetime of clouds measured in millions of years, and its size is a few hundreds of AU, much smaller than the cloud sizes extending on a parsec scale. We may simulate a perturbation by allowing the gas temperature to jump abruptly to higher values, typically a few thousand degrees. Once the warm phase has vanished, the gas cools down and the chemical signatures imprinted by the active stage may persist for a while. The gas then merges back into the cold background, during a longer-lasting relaxation period. Additional parameters are thus the duration of the perturbation, and the peak temperature reached in the warm phase. This description neglects many aspects of the dynamics of perturbation that may significantly affect the chemistry. Nevertheless, even if extremely simplified, such a description retains the global response of the gas to impulsive perturbations. We use the simplified model described above to follow the chemistry during the perturbation. In this specific example, we set the onset of perturbation at $t = 70\,000$ years (close to the final stage of chemical evolution), and a peak temperature $T = 2000$ K lasting for two thousand years. The evolution of a few important species such as CH, CH^+, HCO^+, and CO are reported in Figure 5.8, while Table 5.4 contains the abundances reached at the time the perturbation switches off.

In diffuse interstellar gas, CO is expected to be destroyed by photodissociation and formed by the dissociative recombination of HCO^+

124   Chapter 5

Table 5.4 steady state and end-of-perturbation abundances (cm^{-3}); parameters of the background cloud are reported in Table 5.3; the perturbation triggers at t = 70 0000 years, enhances the temperature from 100 to 2000 K, and lasts for 2000 years.

H	9.93(+1)[a]	9.92(+1)	He	9.99(0)	9.99(0)	C	8.99(−5)	7.38(−5)	O	1.99(−2)	1.98(−2)
H$^+$	4.67(−3)	4.25(−3)	He$^+$	3.74(−4)	3.52(−4)	C$^+$	9.91(−3)	9.88(−3)	O$^+$	1.32(−7)	7.59(−7)
H$^-$	6.24(−10)	7.84(−9)	C$^-$	1.97(−14)	1.62(−14)	O$^-$	2.94(−12)	2.20(−12)	H$_2$	3.50(−1)	3.49(−1)
H$_2^+$	7.82(−11)	1.87(−10)	CH	9.19(−11)	6.24(−9)	CH$^+$	2.13(−10)	4.48(−7)	OH	9.21(−8)	1.23(−4)
OH$^+$	5.94(−8)	2.71(−6)	CO	1.21(−7)	4.96(−5)	CO$^+$	8.82(−12)	1.24(−8)	C$_2$	4.22(−13)	2.06(−10)
C$_2^+$	4.68(−14)	3.30(−11)	O$_2$	4.77(−10)	1.49(−7)	O$_2^+$	1.35(−10)	5.30(−8)	OH$^-$	1.02(−16)	3.94(−16)
H$_3^+$	3.16(−11)	3.52(−10)	CH$_2$	3.03(−12)	7.72(−10)	CH$_2^+$	9.52(−11)	2.97(−8)	H$_2$O	6.57(−10)	2.36(−6)
H$_2$O$^+$	1.85(−9)	3.89(−7)	C$_2$H	8.91(−15)	2.89(−11)	C$_2$H$^+$	4.61(−15)	1.53(−11)	HCO$^+$	1.26(−12)	2.64(−8)
CH$_3^+$	5.64(−12)	6.17(−9)	H$_3$O$^+$	3.70(−11)	3.58(−8)	C$_2$H$_2^+$	2.54(−16)	3.85(−12)	e$^-$	1.49(−2)	1.45(−2)

[a]9.93(1) = 9.93 × 10^1

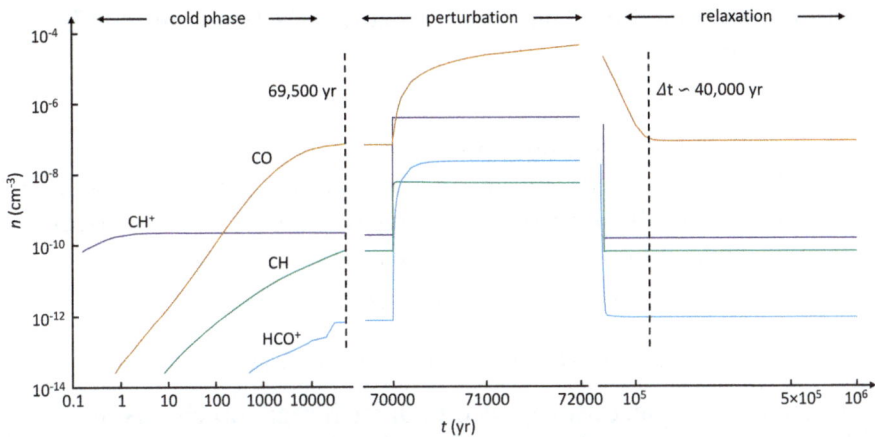

Figure 5.8 Computed abundances of CH, CH⁺, CO, and HCO⁺ in the diffuse cloud model in which a temperature perturbation occurs after 70 000 y of chemical evolution. The perturbation lasts for 2000 y and the temperature rise in the perturbation is from 100 K to 2000 K. the chemical data used in this model are from ref. 6.

(see Section 4.3), regardless of turbulent energy dissipation. In cold chemistry, the formation of HCO⁺ occurs *via* an oxygen hydrogenation chain followed by the two reactions

$$OH + C^+ \rightarrow CO^+ + H$$

and

$$H_2O + C^+ \rightarrow HCO^+ + H.$$

In the process, the conversion of O into O⁺ *via* charge exchange with ionised hydrogen is endothermic by ~230 K. The abundance of HCO⁺ depends on that of H⁺, the production of which is driven by the cosmic ray ionisation of H and H_2. During the perturbation, production of the very reactive cation CH_3^+ is enhanced (see Table 5.4), and this species readily reacts with O to form CO⁺ and HCO⁺. With the opening of this additional formation channel, the abundance of HCO⁺ rises by several orders of magnitude and that of CO by a factor of up to 100 (Figure 5.8). As a consequence, the important hydrogenation chain is not that of oxygen, but that of carbon. As we have seen in Chapter 4, carbon hydrogenation is also at the base of the formation of CH⁺. In the cold chemistry scheme, the destruction of CH⁺ occurs through hydrogenation, an exothermic process with a very short timescale, ~0.45 × $(100 \text{ cm}^{-3})/n_H$ years (from KIDA database). As we have seen in Chapter 4, without sources of suprathermal energy to activate the endothermic reaction (5.43), the formation of CH^+ is initiated by the radiative association

$$C^+ + H_2 \rightarrow CH_2^+ + hv$$

a process with a very long timescale.

Finally, our very small chemical network reproduces well the chemistry of oxygen in the perturbation, occurring along a number of endothermic steps, smoothed by the increase in temperature:

$$O\,(3241\ K) \rightarrow OH\,(1751\ K) \rightarrow H_2O$$

$$O\,(227\ K) \rightarrow O^+ \rightarrow OH^+ \rightarrow H_2O^+ \rightarrow H_3O^+$$

In the cold phase, only the second of such reaction chains proceeds (slowly), until eventually dissociative electron recombination of H_3O^+ gives rise to H_2O, OH, and O.

Some species including CO are not instantaneously removed by the gas phase as a turbulent dissipation phase switches off, but decay to their equilibrium values in the cold gas with a characteristic timescale that is generally much longer than the temporal extension of the perturbation (see Figure 5.8). If the continuous switch on and off of the hot regions in an otherwise cold gas is fast enough, the cloud will be rapidly and densely populated by their chemical relics. Thus, CO traces the relaxation phase, which provides abundances lower than in active regions, but with a much larger filling factor. The relaxation phase has a typical lifetime of roughly 50 000 years, widely independent of the perturbation lifetime.

As a final application of the code, we consider translucent clouds. These regions have typical central extinctions up to $A_V \sim 3$ mag, and they are slightly colder and denser than diffuse clouds. We adopt $n_H = 500$ cm^{-3}, and $T = 30$ K. The size of the cloud is therefore $L \sim 7$ pc. The parameters used to simulate such a translucent cloud are reported in Table 5.4. With these physical conditions, hydrogen is expected to be almost fully converted into its molecular form, and CO will start to incorporate most of the carbon. Since the freeze-out time scales with the gas density are $t \sim 10^9/n_H$ years, some of the gas-phase species might begin to be removed, accreting onto dust surfaces over times of around a couple of million years. The chemical model needs, thus, to be supplemented by a term describing the accretion rate on dust surfaces, see eqn (5.27). We adopt a constant sticking efficiency $S = 0.5$ for neutral and cations. Anions freeze-out more slowly with $S = 0.1$. A sketch of the ongoing chemistry is reported in Figure 5.9. The condensation of gas-phase species onto dust grain surfaces gets substantial for times longer than one million years. Hydrogen is

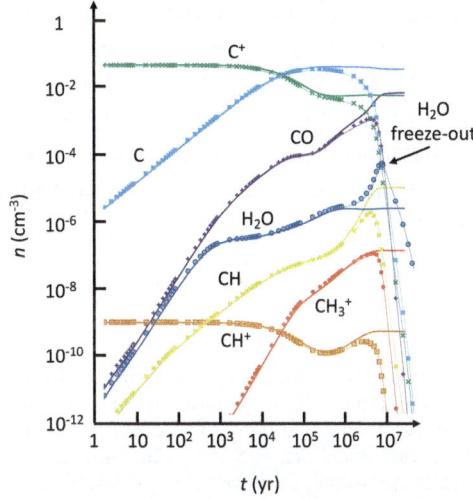

Figure 5.9 Selected computed abundances in the translucent cloud model. In this model, species can be allowed to collide and stick to grain surfaces with defined sticking probabilities. The figure shows both sticking and non-sticking cases. In the former, the loss of material from the gas is dramatic after a few million years of chemical evolution. The chemical data used in this model are from ref. 6.

mostly converted into its molecular form, while carbon is still mainly atomic, although CO has number densities comparable to those of C. Generally, for times longer than the freeze-out time, heavier species begin to disappear from the gas-phase. However, as shown in Figure 5.9, water abundance bucks the trends, and initially starts to increase, because of the increase in its formation rate due to the rise of the ion H_3O^+ abundance. In turn, the concentration of H_3O^+ increases following the decrease of electrons, which occurs because of the condensation of electron providers, such as atomic carbon. As the chemical evolution proceeds, the electron abundance returns to the decaying initial value, due to the freeze-out of most chemical electron sinks.

Translucent clouds are intermediate between diffuse and dense clouds, and therefore show mixed chemical features. In the case presented here, a cloud with moderate density and $A_V = 3$, the gas-phase species are removed on a timescale of the order of a few million years. This visual extinction appears to be close to the critical threshold for ice accretion (see Chapter 8). Such a limiting value, A_{Vcrit}, depends on the nature of the cloud and its environment, and may show variations, but as A_V exceeds this critical value, ice is deposited onto the surfaces of interstellar grains. The chemistry that occurs in such clouds consists of both gas-phase processes and reactions on the surfaces of dust grains, the latter particularly on and in water-dominated ice mantles. It is the interplay between these two types of chemistry that leads to the rich organic chemistry observed in space (see the Appendix to Chapter 1, species composed of more than 5 atoms). We shall discuss this topic in Chapter 8.

References

1. S. S. Prasad and W. T. Huntress Jr., *Astrophys. J., Suppl.*, 1980, **43**, 1.
2. D. A. Williams, T. W. Hartquist, J. M. C. Rawlings, C. Cecchi-Pestellini and S. Viti, *Dynamical Astrochemistry*, Royal Society of Chemistry Publishing, 2018.
3. W. Klemperer, *Proc. Natl. Acad. Sci. U. S. A.*, 2006, **133**, 232.
4. K. Hindmarsh and D. A. House, *J. Chem. Educ.*, 1996, **73**, 585.
5. G. Bussi and A. Laio, *Nat. Rev. Phys.*, 2020, **2**, 200.
6. http://udfa.ajmarkwick.net.
7. D. McElroy, C. Walsh, A. J. Markwick, M. A. Cordiner, K. Smith and T. J. Millar, *Astron. Astrophys.*, 2013, **550**, A36.
8. http://kida.astrophy.u-bordeaux.fr.
9. C. M. Leung, E. Herbst and W. F. Huebner, *Astrophys. J., Suppl.*, 1984, **56**, 231.
10. R. T. Garrod, S. L. Widicus Weaver and E. Herbst, *Astrophys. J.*, 2008, **682**, 283.
11. http://garfield.chem.elte.hu/Burcat/burcat.html.
12. https://www.nist.gov.
13. https://webbook.nist.gov.
14. https://cccbdb.nist.gov.
15. https://jpldataeval.jpl.nasa.gov/index.html.
16. https://phidrates.space.swri.edu.
17. D. A. Verner, G. J. Ferland, K. T. Korista and D. G. Yakovlev, *Astrophys. J.*, 1996, **465**, 487.
18. https://computing.llnl.gov/projects/odepack.
19. D. Galli and F. Palla, *Annu. Rev. Astron. Astrophys.*, 2013, **51**, 163.

6 Chemistry and Dust Formation in Circumstellar Regions and Supernovae

"The dusty stellar atmospheres and envelopes under consideration have a rather complex chemical structure, where, in addition to atoms and ions, a rich zoo of molecules and characteristic dust components also contributes to the emergent spectrum." (H.-P. Gail and E. Sedlmayer)[1]

"Supernovae are considered as prime sources of dust in space. Observations of local supernovae over the past couple of decades have detected the presence of dust in supernova ejecta." (A. Sarangi, M. Matsuura, E. R. Micelotta)[2]

6.1 Introduction

A very crude calculation suggests that for cosmic gas at the average number density of the interstellar medium of the Milky Way (about 1 H atom cm^{-3}) at a temperature of about 100 K, it would take approximately a billion years to accumulate the number of heavy atoms

Astrochemistry: Chemistry in Interstellar and Circumstellar Space
By David A. Williams and Cesare Cecchi-Pestellini
© David A. Williams and Cesare Cecchi-Pestellini 2023
Published by the Royal Society of Chemistry, www.rsc.org

(such as Fe, Si, or Mg) that are expected to constitute a large interstellar silicate grain (of material such as forsterite, Mg_2SiO_4, or fayalite, Fe_2SiO_4) of radius 0.1 µm. A billion years is a very long time; in fact, it means that dust formation in this way would be far too slow to account for the dust that is present and detectable in the Milky Way. Evidently, interstellar dust cannot be formed easily in the very low density gas that fills interstellar space.

Nevertheless, it's a remarkable fact – as we shall describe in this chapter – that we can observe from terrestrial observatories, *in real time*, the effects of dust being formed in various particular locations in the Milky Way galaxy. The timescales for dust formation are observed to be short, even by human standards. These locations include the ejecta of supernovae, where – if dust appears – it seems to be formed within a year or so after the supernova explosion. Dust is also present in the extended envelopes of some stars of about solar mass near the ends of their lives; in situations in which dust is present in these envelopes, it is observed in these examples to form very rapidly, perhaps even on timescales as short as a month or so.

The gas number densities in the ejecta of supernovae and in stellar envelopes are very much greater than the average number density of the interstellar gas in the Milky Way. So perhaps it is not surprising that the dust formation timescale in those locations is observed to be very much shorter than the very crude estimate of about a billion years for dust formation in the low density gas in interstellar space. In our study, we shall find the chemistry that is a precursor event to dust formation.

This chapter is devoted primarily to a discussion of dust formation in these two locations: circumstellar envelopes and supernovae ejecta. We'll discuss the formation of dust in stellar envelopes in Section 6.2 and the formation of dust in supernovae ejecta in Section 6.3. There are also other interstellar locations where dust grains are seen to form, and we shall list them and their potential contributions to dust formation in the final section of this chapter, Section 6.4. It is important to distinguish between *circumstellar* dust, *i.e.*, dust formed in circumstellar regions, and *interstellar* dust, *i.e.*, dust that is widespread throughout the interstellar environment (as discussed in Chapter 1, Section 1.1.4). Therefore, we shall also discuss in Section 6.4 what happens to dust formed in both these types of circumstellar region after it has been ejected into interstellar space.

6.2 Formation of Dust in Circumstellar Envelopes and Its Ejection into Interstellar Space

6.2.1 Dust in Circumstellar Envelopes: Some Background Information

While most stars shine steadily without any apparent change during human lifetimes, some stars show remarkable variability. These variable stars inevitably attract attention: why do some of them apparently switch on and off? It can hardly be due to the energy source that powers the stars themselves. A very early study, in 1783, by Edward Pigott, of a star in the constellation of Coronae Borealis showed that the brightness of this star (now called R CrB) decreased within a period of a month so that the star eventually could not be seen.[3] Effectively, the star had disappeared, only to reappear within about a year and a half, following which the star's brightness remained constant for about ten years. This type of disappearing trick was then repeated at irregular intervals, and is still occurring. Later spectroscopic work showed that the atmosphere of this star was rich in carbon and helium but poor in hydrogen, so it was suggested that episodic formation of carbon-based soot might cause enough extinction to obscure the star's light totally. The eventual dissipation of the dust (or soot) into interstellar space would then allow the perceived brightness of the starlight to recover so that the sequence of events could be repeated. This interpretation of the behaviour of R CrB is now accepted. There are other stars closely similar to R CrB showing the same general behaviour. However, the number of stars in this particular class is very small and their dust contribution could not account for the amount of dust in the Milky Way.

The main problems raised by this interpretation of the behaviour of R CrB included, firstly, the nucleation and growth of solid particles in the environment of the star, and, secondly, the ejection and dissipation of the sooty gas into interstellar space. More detailed studies of these problems in a variety of stars – other than R CrB stars – showed that dust formation was likely to occur not only in carbon-rich stellar envelopes (*i.e.*, stars with more carbon atoms than oxygen atoms in their atmospheres) where it was plausible to expect that carbon soot might appear, but also in oxygen-rich envelopes (*i.e.*, with more oxygen than carbon). In oxygen-rich stellar atmospheres one might expect solids such as silicates and oxides of various kinds to form, rather than soot.

In both cases, it seemed likely that radiation pressure from the star on dust grains might be able to drive the grains away from the star and into interstellar space. Collisions between these driven grains and gas molecules couple the motion of the grains to that of the gas, and so a dusty wind flowing from star into space might be created.

During the last half-century, infrared observations have made enormous contributions to astronomy. They have confirmed that dust exists in the circumstellar regions of many types of star, not just the circumstellar regions of the R CrB stars. Infrared observations reveal not only the presence of circumstellar dust by the broad-band extinction it causes if the dust is cool or continuum emission if the dust is warm, but also much spectral information that can reveal its composition. These infrared observations confirm the presence in circumstellar dust of various kinds of silicates, some kinds of carbons, and some other materials. For example, silicates show two strong infrared features, one broad feature centred at a wavelength of 9.7 µm arising from the Si–O stretching vibrations in the SiO_4 tetrahedron, and another broad feature centred at about 18 µm arising from the O–Si–O bending vibrations in the SiO_4 tetrahedron. Solid carbon containing benzene-type rings (sp^2 bonding) with attached hydrogen atoms shows features arising from the stretching (3.3 µm) and bending vibrations of the C–H bond in-plane (8.6 µm) and out-of-plane (11.3 µm) at the edge of the ring structure. Stretching vibrations of the C–C bond in the benzene-type ring also contributes a feature at 6.2 µm. Aliphatic carbon (sp^3 bonding) provides a C–H feature at 3.4 µm. A very detailed account of the physics and chemistry of circumstellar dust shells can be found in ref. 1.

Besides the (fairly rare) R CrB stars, the presence of dust in circumstellar regions has been confirmed in many types of (quite common) stars, so circumstellar dust is not a minor phenomenon but is widespread in the galaxy (and also in external galaxies). Many of these stellar types have cool extended envelopes: these include stars known as red giants, supergiants, the so-called asymptotic giant branch (AGB) pulsating stars, and also protoplanetary nebulae and planetary nebulae, and OH/IR objects. The great astronomical narrative describing the formation of stars from interstellar gas and the evolution of these stars as they come towards the end of their lives is described in Section 6.2.2 and is illustrated in the *Hertzsprung–Russell diagram*, see Figure 6.1. Some readers may wish to omit the astronomical details in Section 6.2.2, and for them the essential conclusions of that Section are summarized in Table 6.1. A list of some objects containing circumstellar dust can be found in Table 6.2.

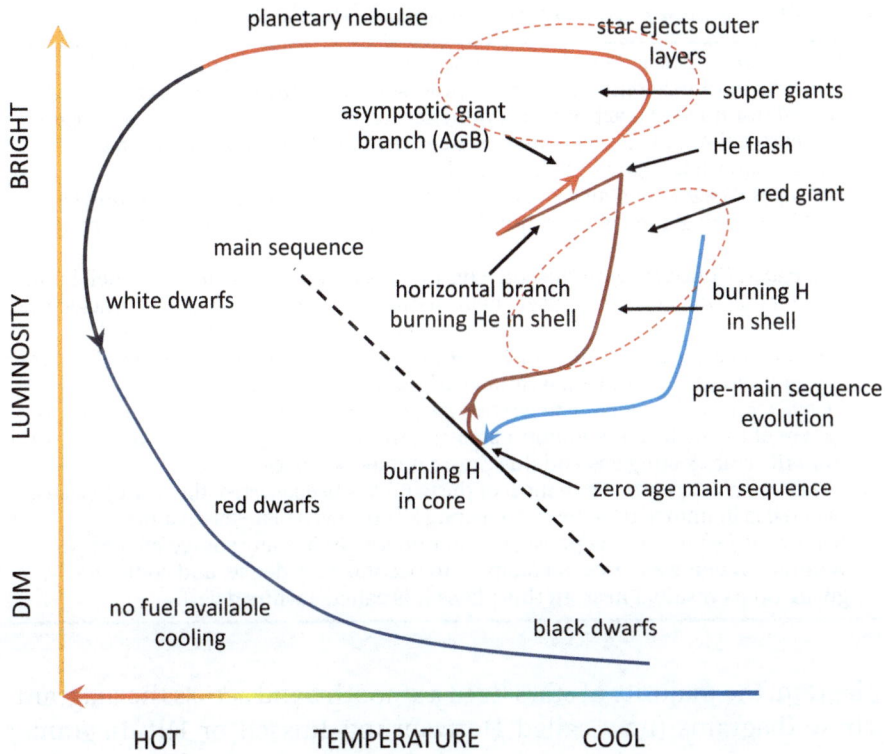

Figure 6.1 A schematic HR diagram, illustrating the evolution of a cloud of gas and dust into a star of about a solar mass, and of the star's subsequent passage through various stages. The evolution begins in the pre-Main Sequence phase (light-blue line) until the star arrives on the Main Sequence (black diagonal line). On the Main Sequence, the star burns hydrogen in its core and remains for most of its life. The star then leaves the Main Sequence (following a path indicated roughly by the brown line), passing relatively rapidly through various phases (red and purple lines) including the AGB and planetary nebula phases, through the white dwarf phase, eventually becoming a black dwarf.

6.2.2   Stellar Evolution

On human timescales, most stars appear to be immutable. However, all stars change in time, and can change very much during their lives, from when they are formed until they run out of energy. The description of the way stars change with time is called stellar evolution.

At the beginning of the last century, the Danish Astronomer Einar Hertzsprung, and – independently – Henry Norris Russell, noticed that when the luminosity and temperature of stars are plotted in a

Table 6.1 Summary: the evolution of a solar mass star.

1. *Main Sequence phase*: The material that forms the star is initially almost all hydrogen. Gravitational contraction of the protostar raises the core temperature high enough for nuclear reactions to commence, converting almost all its hydrogen nuclei to helium nuclei. This main sequence phase lasts about a billion years. The star now has a core of helium nuclei.
2. *Red Giant phase*: When almost all the hydrogen is used up, the star expands and cools, becoming a red giant, a hundred times larger than the Sun in its previous phase.
3. *AGB phase*: Contraction in the core compresses and heats the helium nuclei until nuclear reactions rapidly convert helium nuclei to carbon and oxygen nuclei; this causes the "helium flash". The star is now called an AGB star. Its atmosphere is cool, and is extended by stellar pulsations. The atmosphere may be O-rich (more oxygen than carbon) or C-rich (more carbon than oxygen).
4. *Chemistry, dust formation, and ejection phase*: these processes occur in the atmosphere of the AGB star. Radiation pressure from the star on the dust drives a wind from the star ejecting gas and dust into interstellar space.
5. *Planetary nebula phase*: When most of the wind has been ejected, the nearby ejected material is illuminated by the star, creating a short-lived planetary nebula.
6. *White dwarf phase*: In this phase, the star no longer has any source of energy. Without an energy source, it collapses to become very dense, and cools slowly, giving up its residual heat. In this phase it is called a white dwarf.

diagram, the majority of stars lie in a smooth band across the diagram. These diagrams (now called Hertzsprung–Russell or HR diagrams) are one of the most important tools in the study of stellar evolution. How long a star lives and what its ultimate fate may be depend on how much mass it has when it is formed. The initial mass of a star determines its evolutionary stages, with massive stars evolving much more rapidly than low-mass stars. Each new evolutionary stage corresponds to a change in luminosity and temperature of the star, and therefore to a new position in the HR diagram, so stellar evolution implies a track in the HR diagram (see Figure 6.1).

The majority of stars, about 80–90% of all the stars (including the Sun), are found to occupy a region in the HR diagram called the *Main Sequence*, running from top left (hot, bright stars) to bottom right (cool, dim stars). The luminosity of a star increases with its size, since larger stars have greater surface area through which radiation can flow. This explains the presence of stars above the Main Sequence with the same temperature as cooler Main Sequence stars, but having greater surface areas. Stars that have the same luminosity as dimmer Main Sequence stars are hotter (lying to the left of them on the HR diagram) and have smaller surface areas. These are termed *dwarf* stars. The converse is also true, and the larger stars are called *giants*.

Table 6.2 List of some types of star that have circumstellar dust. The Table shows that dust formation is a common property of stars in the Milky Way galaxy, and – by implication – the Universe.

AGB star: An AGB star is a cool, luminous ($L \sim 10^3 \, L_\odot$) red giant star of mass (0.6–10) $M_\odot$ in a late stage of evolution. The star may be pulsating. It has an extended atmosphere. The pulsations and radiation pressure on dust formed in the atmosphere generate a wind, leading to significant mass loss (up to $\sim 10^{-4} \, M_\odot \, y^{-1}$). These stars are important contributors of dust in the Milky Way.

Planetary nebula (PN) and protoplanetary nebula (PPN): A PN is an emission nebula consisting of an expanding shell of ionized gas, originally ejected from and illuminated by an AGB star in a late stage of evolution. A PPN is a brief phase in the transformation from AGB to PN.

S and C classification of stars: These are the coolest classes of stars. A class C star is an AGB star with more carbon than oxygen in its atmosphere.

OH/IR star: An AGB or red supergiant star showing strong OH maser emission at 18 cm; the OH derives from H_2O formed in the AGB atmosphere. The star is unusually bright at infrared wavelengths because of emission from heated silicate dust formed in the AGB atmosphere.

Herbig Ae/Be star: This is a pre-main sequence star (*i.e.*, a protostar heated by gravitational contraction rather than nucleosynthesis), usually embedded in a cloud of gas and dust, with a circumstellar disc, winds and (sometimes) jets. Although these are dusty objects, their dust is interstellar, and has not been formed *in situ*.

Wolf–Rayet (WR) star: A WR star is a normal stage of evolution of a very massive star. If broad lines of N and C are in the spectrum, the star is further classified as WN or WC, respectively. WR stars have very powerful winds. WC stars produce dust, often in collisions of winds in binary WCs, confirming that dust can be formed in extreme environments.

Luminous Blue Variable (LBV): A LBV is a massive evolved star with erratic and violent variations in its spectrum and luminosity. At least one example (AG Car) shows an infrared continuous spectrum confirming the presence of large dust grains and a very high dust mass.

We can plot a "track" on an HR diagram that represents how the temperature and luminosity of a star change over time. The star moves in the diagram along an evolutionary path, in response to contraction under gravity and the rate and the fuel of nuclear burning. Newly-forming stars change in size because they are contracting under gravity, while Main Sequence stars change because they are using up their nuclear fuel. During their evolution, some stars may pass through a phase in which they are not in equilibrium and they pulsate, expanding and shrinking in response to internal forces including pressure and gravity. This pulsation drives changes in the luminosity. During this stage, stars cross an area called the Instability Strip on the HR diagram, a narrow almost vertical region located between mid-sized stars on the Main Sequence and the Giant Branch. Cepheid Variables are an important type of pulsating stars. These Cepheid Variables are used as distance indicators because the period of their pulsations varies in proportion with their luminosity. From the period, we obtain the true luminosity, and comparing with the actual luminosity tells us the distance to the star. Cepheids are near the top of the Instability Strip.

To estimate how the luminosity and temperature of a star change as it ages, astronomers must resort to calculations. A model predicts the luminosity and size of the star, and from these values we can determine its surface temperature. For instance, a star like the Sun is initially embedded in a cloud of gas and dust and is not visible until it is outside the cloud. At this stage, it is better to call it a *protostar*. Its internal temperature is about 1 000 000 K but the outer layers of this protostar are much cooler, with a temperature of 3500 K. As the protostar contracts under gravity, the temperature of the outer layers increases, but the net effect is a decrease in luminosity so that its representative point in the HR diagram moves down and to the left. This process will continue until it reaches the location in the HR diagram corresponding to 0.7 solar luminosities and 4500 K. Finally, when the star is burning hydrogen in its core through fusion reactions and has reached equilibrium, it will lie on the Main Sequence; it will have a surface temperature of about 6000 K and a luminosity corresponding to the present Sun. The time when the star first joins the Main Sequence on the HR diagram is called the Zero Age Main Sequence. Once the star reaches equilibrium and until it exhausts its hydrogen fuel, it stays on the Main Sequence.

A star more massive than the Sun has a hotter core because – thanks to stronger gravity – it can reach higher pressures before they in turn generate enough radiation to slow down the contraction. Thus, more massive stars produce energy at a much faster rate than low-mass stars, and are brighter, but they live for a much shorter time. Massive stars stay on the Main Sequence for about 10 *million* years, while a star like the Sun has a lifetime of about 10 *billion* years. The Main Sequence lifetime of cooler stars may reach the astounding age of 10 *trillion* years! This is considerably longer than the current age of the Universe.

What happens when a Main Sequence star exhausts the hydrogen in its core? With no additional fuel in the core, the nuclear fusion converting hydrogen to helium dies out. The core loses the battle against gravity and contracts. As it shrinks, it heats up. This increase in temperature is enough to ignite the fusion of hydrogen embedded in the exhausted core. The newly increased radiation pressure actually causes the outer layers of the star to expand. As the gas expands, it cools, which causes the effective temperature to drop. During this stage of expansion, the star will leave the Main Sequence, moving up (to higher luminosities) and to the right (to lower temperatures) along the so-called Red Giant Branch. The Sun will become a larger but cooler star called a *red giant*. It will be 100 times bigger and its temperature will be about 3500 K.

In the meantime, the helium core – left over from the hydrogen burning phase – contracts until its temperature rises to over one hundred million degrees. At this point, new nuclear reactions set in, converting helium nuclei to carbon and oxygen nuclei. The star expands again, but not enough to compensate for the increased energy generation, so the temperature in the core increases without restriction. This occurs because when a gas becomes super-compressed (the technical term is *degenerate*), it behaves more like a solid. If a normal gas is heated, it expands. However, the pressure in a degenerate gas does not depend on the temperature. Because the expansion does not compensate, temperatures remain very high and the helium burning occurs quickly and is uncontrolled. This sudden onset of helium core fusion is called the *helium flash*. During the helium flash the stellar luminosity remains constant. Thus, since the temperature is increasing the star's position in the HR diagram moves to the left, roughly horizontally, so stars in this phase are said to be on the Horizontal Giant Branch.

At the end of the helium burning phase, the core contracts, but to no avail. The end products of helium burning – carbon and oxygen – are unable to ignite their nuclear fusion. However, the core contraction generates sufficient heat for the surrounding layer of helium to start fusing, which in turn heats up surrounding unused hydrogen and this starts to burn. The giant star expands again, reaching a size as large as the orbit of Mars and the luminosity increases. The star moves along a track that lies above and roughly parallel to the Red Giant Branch, and for that reason this track is called the Asymptotic Giant Branch (AGB). In this phase, dust grains are formed in the upper atmospheric layers of the star. Eventually, a wind develops in the star's envelope, blowing the outer layers into interstellar space. When the envelope of the star is nearly gone, the star becomes a *planetary nebula*. This consists of an expanding glowing shell of ionized gas. These objects owe their name to a misunderstanding by William Herschel: he saw that these objects appeared to be round, like planets.

The exposed remnant core that ionized the planetary nebula material is basically an extremely hot dense sphere of carbon and oxygen, a so-called *white dwarf* with a surface temperature of about 10 000 K. No more nuclear fusion is possible in a white dwarf star, and no further gravitational collapse can occur, so energy generation ceases.

The relic star radiates its energy away and slowly fades from view, becoming a *black dwarf*.

The description of stellar evolution given here is for stars with masses in the approximate range of 0.8–8 solar masses. Stars with greater masses are unlikely to survive as red giants. Instead, they will destroy themselves as supernovae (which we'll discuss in Section 6.3). Table 6.1 summarizes this description.

The zoo of objects that show the presence of circumstellar dust is large, and the astronomical nomenclature is often imaginative but uninformative. For convenience, we show in Table 6.2 a list of the most important objects of this type, and give a few words about their origin. Figure 6.2 shows an image of a star showing repeated episodes of dust formation and ejection, and an image of a planetary nebula into which these stars evolve on their paths towards white dwarf and black dwarf status.

6.2.3   O-Rich and C-Rich Atmospheres of AGB Stars

We have stressed already that it is necessary to know whether the total abundance of oxygen in a stellar atmosphere exceeds that of carbon (O-rich) or, conversely, the total abundance of carbon exceeds that of

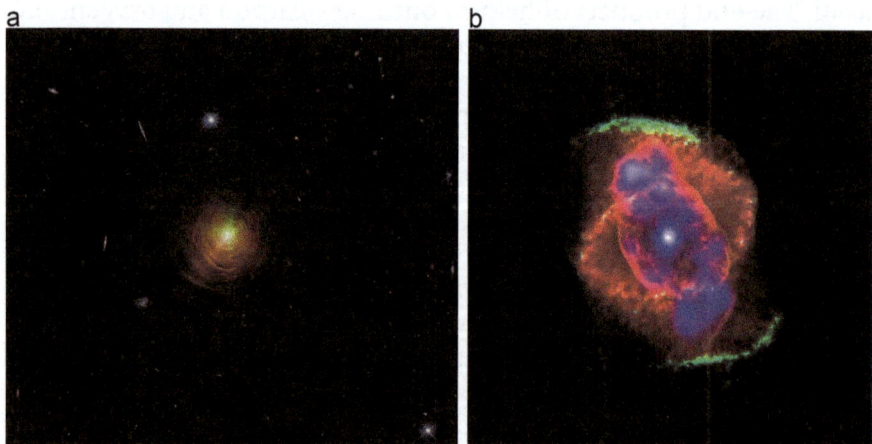

Figure 6.2   (a) An image of a star showing repeated episodes of dust formation and ejection. Many episodes of dust formation are seen in this Hubble Space Telescope image of IRC+10216 (also known as CW Leonis), a carbon-rich star with a thick dusty envelope (credit ESA/Hubble & NASA, T. Ueta, H. Kim). (b) Complex modes of ejection are seen in the Cat's Eye planetary nebula. [credit: NASA/X-ray: Y. Chu (UIUC) *et al.*, Optical: J. Harrington, K. J. Borkowski (UMD), Composite: Z. Levay (STScI)].

oxygen (C-rich). These ratios determine the nature of the chemistry and dust that may form in these regions. We should remember that the gas is, of course, totally dominated by hydrogen, by a factor of more than one thousand, so these terms, O-rich and C-rich, apply only to these two elements, oxygen and carbon. The fundamental reason for the distinction between these two types of circumstellar region is that carbon monoxide, CO, is one of the most stable diatomic molecules. When formed in circumstellar gas, CO takes up the maximum possible amount of the carbon and oxygen available. Only the excess of the more abundant element, either O or C, takes part in further chemistry. As we'll see, in O-rich gas, there is obviously excess oxygen after CO forms, and this oxygen forms a variety of oxides, while in C-rich gas there is excess carbon, which forms a variety of carbons and hydrocarbons. To evaluate the chemical equilibrium in AGB star atmospheres (these are important locations for dust formation), we need to know the elemental composition before the chemistry begins, especially whether it is O-rich or C-rich. In this section, we describe more details about the elemental composition of AGB stars, and how this composition evolves. A summary of our conclusions is given in Table 6.3.

As we have seen in the preceding section, the nuclear energy sources for stars that become AGB stars are, firstly, hydrogen-burning (to produce helium nuclei) and, secondly, helium-burning (to produce carbon, nitrogen, and oxygen nuclei in particular, and other nuclei too). The precise evolutionary path of a star is determined by the stellar mass. For stellar masses greater than 0.5 solar masses, both these energy sources are involved. If the stellar mass is smaller than this limit, then no helium-burning occurs and so there is no formation of C, O, and N. The elements C and O are those that are most likely to be involved in dust formation, and so no dust formation is possible for these very low mass stars.

Hydrogen-burning proceeds most rapidly at the centre of the star where the temperature is highest. Therefore, a helium core develops and grows as the hydrogen is converted to helium, and the helium core is surrounded by a hydrogen-burning shell. The helium core contracts and becomes denser and hotter while the envelope expands. Eventually, the helium core ignites, the details depending sensitively on the stellar mass. The important point for present considerations is that a carbon–oxygen core develops inside the helium core through reactions in which three ^{4}He nuclei (or α-particles) are converted to a ^{12}C nucleus, while a ^{12}C nucleus and one α-particle create a ^{16}O nucleus. Reactions of these nuclei with protons, α-particles, electrons and positrons, γ-rays and neutrinos in the so-called CNO cycle generate other isotopes of carbon (^{13}C) and of oxygen (^{15}O, ^{17}O, and ^{18}O), isotopes of nitrogen (^{13}N, ^{14}N, and ^{15}N), and some other elements. The important balance between oxygen and carbon in the core is controlled by the stellar mass because it determines the temperature.

Although this carbon–oxygen core is deep inside the star, a convection zone in which turbulent mixing is occurring ensures that abundance gradients are removed. Convection is particularly important in stars of low and intermediate mass; in low mass stars the convective zone may occupy the entire bulk of the star, while in a star like the Sun convection affects the material lying in about the outer third of a radius. In AGB stars, however, the convection zone can be deep, so that material (enriched in C, O, and N, especially, but also Mg, Si, and Fe) from the central core can reach the stellar atmosphere. This process is called "dredge-up", and the evidence supporting it lies in the isotopic anomalies arising in the helium core chemistry that are detected in presolar grains (*i.e.*, grains formed before the formation of the Sun) located in meteorites. Dredge-up may occur a number of times during the life of an AGB star.

After the helium shell has ignited and the helium fuel begins to be consumed, helium-burning declines. Hydrogen-burning in a shell surrounding the helium shell then resumes, introducing some restructuring in which the core becomes smaller and hotter. Eventually, helium-burning may be able to resume, until – again – much of the helium is consumed. Then hydrogen-burning resumes once again and the cycle may repeat. These switches of energy source and consequent restructuring affect the stellar energy output, and such a star is called a *thermally pulsating AGB star*. The pulsation stimulates mass loss through a stellar wind, although radiation pressure on dust grains produced in the envelope of the star may eventually dominate as the driver of the wind. The maximum mass loss in the wind may occur at a very high rate, as high as 10^{-4} $M_\odot$ y^{-1}. Evidently, such a high mass loss rate cannot be sustained for very long.

Detailed models show that the likelihood of an AGB star becoming a carbon-rich star (as opposed to remaining an oxygen-rich star) is strongly dependent on stellar mass. Detailed computations show that for stellar masses less than 1.55 solar masses, stars remain oxygen-rich, but at that mass, a short-lived carbon-rich phase (3.7×10^4 y) can arise, following the oxygen-rich phase whose duration is almost ten times longer. However, for a stellar mass of 2 solar masses, the oxygen-rich and carbon-rich phases have similar durations (at

Table 6.3 Essential summary: O-rich and C-rich AGB stars.

1. *The C:O ratio*: The C:O ratio in AGB stars is determined by the initial mass of the star. However, stars of very low mass (<0.5 solar masses) never reach the helium-burning phase, and therefore contain only H and He. They cannot become AGB stars and they cannot make dust.
2. *Dredge-up*: The atmospheres of AGB stars of a solar mass and greater are affected by convection and turbulent mixing. This mixing (called "dredge-up") brings material (enriched in C, O, and N, and other elements such as Si, Mg, and Fe) from the stellar core to the surface, enriching the atmosphere in these elements. The enrichment of the stellar atmosphere may happen several times during the life of an AGB star.
3. *Effect of stellar masses*: The duration of C-rich phases of an AGB star is highly dependent on the stellar mass. For a mass of 2 solar masses, O-rich and C-rich phases have similar durations ($\sim 3 \times 10^5$ y), while for a 4 solar mass star the C-rich phase is longer than the O-rich phase. For larger than 5 solar masses, no C-rich phase exists.
4. *Excess elements*: In O-rich stars, almost all the carbon is tied up in CO and the excess oxygen appears mainly in oxides and silicates. In C-rich stars, almost all the oxygen is tied up in CO and the excess carbon appears mainly in carbons and hydrocarbons.

around 3.5×10^5 y), while for stellar masses of 3 solar masses (which corresponds to the longest duration as an AGB star), the carbon-rich phase duration ($\sim$1.6 My) is about four times longer than the oxygen-rich phase. At five solar masses and above, AGB stars do not have a carbon-rich phase.

Relative to the Solar System values, carbon and oxygen fractional abundances are slightly reduced in O-rich stars while that of nitrogen is enhanced. In C-rich stars, carbon is strongly enhanced by dredge-up, so that the abundance of carbon exceeds that of oxygen. Typical ratios of O/H and C/H measured in AGB stars are both in the range $(4-8) \times 10^{-4}$. A summary of this rather complex evolution of AGB stars is given in Table 6.3.

6.2.4 Chemistry in AGB Atmospheres

Let's begin by considering the chemistry in thermodynamic equilibrium in the circumstellar gas of AGB stars. This is the gas into which atoms of C, N, O, and other elements have been injected from the stellar interior by dredge-up, and in which solids may begin to condense when the circumstellar gas is set into a cooling outflow from the star by a succession of mild shocks, driven by pulsations observed to occur in the underlying star. Dust formation typically occurs in circumstellar gas for temperatures less than about 2000 K, depending on the nature of the dust. The chemical composition of the circumstellar gas and the local physical conditions in the cooling outflow determine if and how nucleation and growth of solid particles can occur.

To describe chemical equilibrium for the gas phase preceding the dust formation phase, we can write down a set of algebraic equations involving the partial pressures of the free atoms, the free enthalpies of formation of molecules formed from free atoms of all the elements forming those molecules, and the stoichiometric coefficients of elements involved in a particular molecule. The set of equations for the partial pressures of all species in the system can be solved to give the complete chemical composition of the system. We describe how these equations are derived in the following sub-section 6.2.4.1 and show some examples of the results in 6.2.4.2.

6.2.4.1 Setting Up the Equations for Gas Phase Chemistry in Local Thermodynamic Equilibrium

The aim is to solve the equations determining partial pressures for all atoms and molecules in the system. The equations involved are the law of mass action and the conservation of the elements. Obviously, the partial pressures obtained from the solution may be expressed as number densities, if required.

If a molecule ABC... is a molecule formed from N atoms A, B, C, ...

$$A + B + C + ... \rightarrow ABC...$$

then in chemical equilibrium the partial pressure of the molecule is given by the law of mass action

$$p_{ABC...} = p_A \, p_B \, p_C ... \exp(-\Delta G/RT)$$

where ΔG is the change in free enthalpy during the formation of the molecule, so that

$$\Delta G = \Delta H - T\Delta S = \Delta H_f\,(ABC...) - \Delta H_f(A) - \Delta H_f(B) - ... -T\,[S(ABC) - S(A) - S(B) - ...]$$

in the conventional notation of thermodynamics. The free energies of formation may be derived from laboratory experiments, and also from calculations when the required data on energy states and molecular structure are known. The free energies of formation are tabulated in various sources (see ref. 1) and may also be found on web pages. Some of these data are given in the Burcat database mentioned in Chapter 5.

We also have N equations of conservation which constrain the total number of atoms A (or B, C, ...) in all molecules in the stellar atmosphere to be equal to the number of atoms of A (or B, C, ...)

available. The total number of atoms A is related to the total pressure by the elemental abundance of A (or B, C, ...) available in the stellar atmosphere.

In principle, all these equations are together sufficient to allow a solution to be obtained; *i.e.*, to obtain all the partial pressures of all species in the atmosphere. In practice, the equations are complicated and various methods have been introduced to obtain the solutions. One such method is hierarchical, in which the calculation involves approaching the true solution in successive steps. Alternatively, standard procedures such as the Newton-Raphson method can be used.

In the hierarchical method (see ref. 1 for more details), the approach is to focus first on the most abundant molecules, and then introduce molecules of lesser abundance. For illustration, we consider here the hierarchy of elements containing oxygen in an O-rich atmosphere. Let us assume that the most abundant O-containing species in such an atmosphere, before this gas drifts away and cools, are CO, H_2O, and OH. The equation of conservation is, therefore,

$$\varepsilon_O P_H = p_O + p_{CO} + p_{OH} + p_{H2O} + p_{O,min}$$

where P_H is the so-called fictitious pressure calculated assuming that all H atoms, whether as H or H_2, contribute as atoms to the pressure, and ε_O is the fractional abundance of oxygen in the stellar atmosphere relative to hydrogen. On the right hand side are the partial pressures of O, CO, OH and H_2O, and the final term is the partial pressure of the other (less-abundant) oxygen-containing molecules, so that

$$p_{O,min} = p_{SiO} + 2p_{SiO2} + 2p_{CO2} + ...$$

Writing the laws of mass action

$$p_{OH} = p_O p_H K_p(OH) \text{ and } p_{H2O} = p_O p_H^2 K_p(H_2O)$$

where $K_p(OH)$ and $K_p(H_2O)$ are the equilibrium constants which each have the form $\exp(-\Delta G/RT)$ with appropriate ΔG (see above), we can obtain the following expression for the partial pressure p_O:

$$p_O = [\varepsilon_O P_H - p_{CO} - p_{O,min}]/[1 + p_H K_p(OH) + p_H^2 K_p(H_2O)]$$

Since almost all the carbon is tied up in CO, we can write $p_{CO} = \varepsilon_C P_H$. The equation for p_O may be solved iteratively: we can ignore the small term $p_{O,min}$ in the first step and calculate p_O from the above expression. Then the partial pressures of all species containing H and O can be obtained, and in the next iteration a better approximation for all these species can be obtained. The process can be repeated as often as required until the desired accuracy is achieved.

Similar methods can be applied to determining the partial pressures for carbon-containing species, starting with the analogous conservation equation

$$\varepsilon_C P_H = p_C + p_{CO} + p_{CO2} + p_{CH4} + p_{C,min}$$

and proceeding in a similar manner as in the oxygen case. Similarly, partial pressures of species containing the elements nitrogen and silicon can be obtained, and successive iterations give improving values of the partial pressures of all the species in the stellar atmosphere. In this way, a complete solution of the chemistry, to a desired accuracy, may be obtained.

6.2.4.2 Some Results for Equilibrium Chemistry in Cool Stellar Atmospheres

Results (from ref. 1) showing number densities of species in the chemistry in oxygen-rich and carbon-rich cases of AGB stars are shown in Figure 6.3a and b. The variation of the chemistry is shown as a function of temperature from 2000 K to 600 K for gas at a pressure of 10^{-10} bar. These abundances are shown prior to thermal pulsing of the star.

These rather complicated figures allow some quite simple and powerful conclusions to be drawn: they suggest that the growth of dust grains in AGB atmospheres depends on a rather few reasonably abundant atomic and molecular species. In O-rich atmospheres, the main species likely to be important for grain growth are SiO, Mg, Fe, and H_2O, with minor contributions from Ca, Al, AlOH, and Al_2O. In C-rich atmospheres, the growth of carbon grains occurs through C_2H and C_2H_2 and of silicon-bearing grains through Si, Si_2C, and SiS.

6.2.5 Formation of the Nuclei of Dust in AGB Stars

The scheme of dust formation in stellar outflows that is summarized in Figure 6.4 envisages the formation of nuclei which form centres of growth for dust grains. Once the dust grains have formed, they may become an important driving force in maintaining the outward flow of the cooling stellar envelope through the action of stellar radiation pressure.

The first nuclei to condense out as the circumstellar gas moves away from the star and cools down do not necessarily have the same composition as the material that condenses on them during later grain growth, but are formed from the first materials that are able to condense out from the gas, at the highest temperatures during this cooling

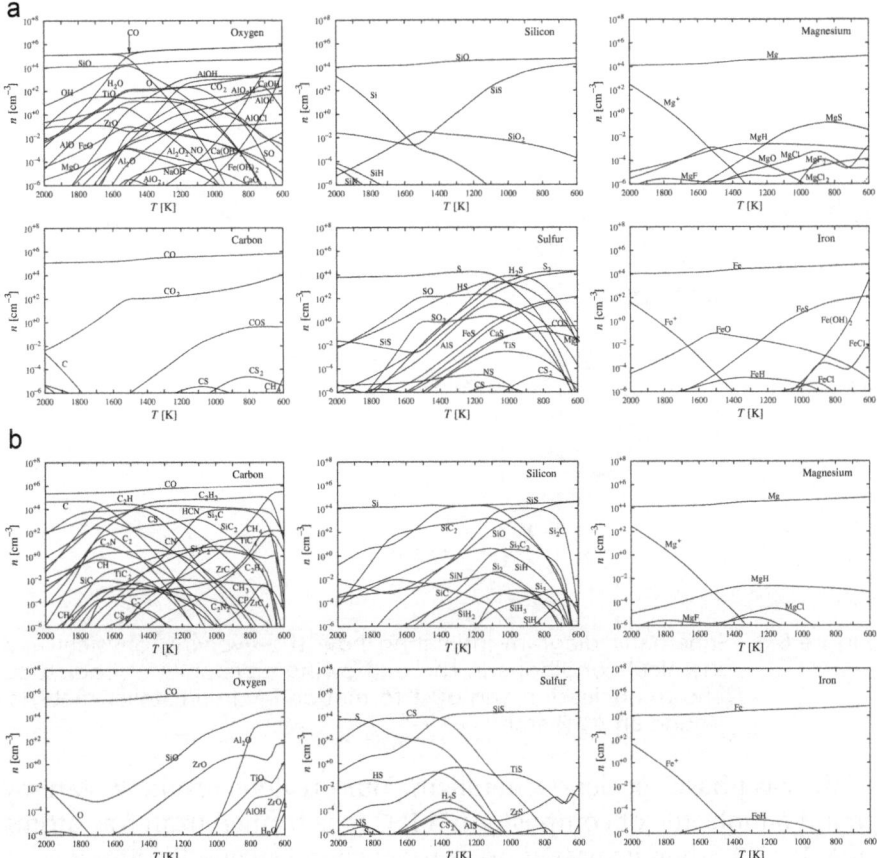

Figure 6.3 (a): Oxygen-rich case: chemistry for oxygen, silicon, magnesium, carbon, sulfur, and iron; (b): carbon-rich case, chemistry for carbon, silicon, magnesium, oxygen, sulfur, and iron. Abundances of atoms, ions and molecules are shown for the oxygen-rich case (a) and carbon-rich case (b) for a fixed pressure of 10^{-10} bar. The abundances are those at the stellar surface before outflow is initiated. The figures are taken with permission from Gail and Sedlmayer.[1]

process. It is necessary, therefore, to consider the chemistry that occurs in the outflow, as described briefly in the previous sub-section, but to include also the possibility that dust grains form in the gas. Such solids may be in equilibrium with the gas, if accretion and evaporation are in balance. If accretion is dominant, then the nuclei grow, but if evaporation is dominant, then the nuclei tend to disappear.

In the discussion presented in subsection 6.2.4.1, the presence of dust was ignored. It can be included[1] by analogy with the discussion

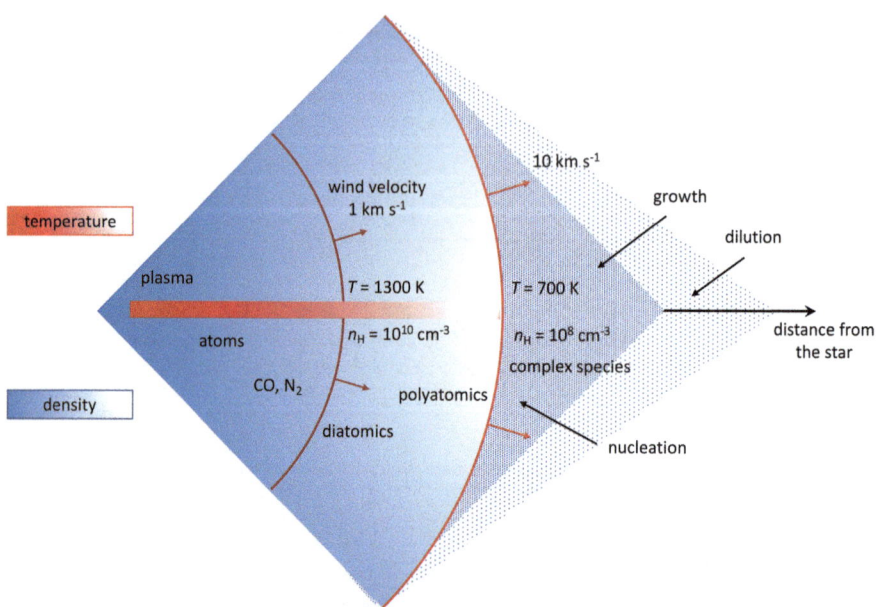

Figure 6.4   Schematic diagram indicating how the evolving physical and chemical conditions in the outflowing envelopes of cool stars lead to nucleation and dust formation in circumstellar material around an AGB star.

of the gas phase species. Generalizing our previous result, we can say that if a molecule of composition $A_iB_jC_k...$ is formed from free atoms of A, B, C, ..., then its partial pressure in chemical equilibrium is

$$p\left(A_iB_jC_k\cdots\right)=p_A^i p_B^j p_C^k\cdots\exp(-\Delta G/RT)$$

If a solid with the same composition is formed from gas phase atoms, this solid is in equilibrium with the gas phase if

$$1=a^c\left(A_iB_jC_k...\right)=p_A^i p_B^j p_C^k...\exp(-\Delta G/RT)$$

If $a^c(A_iB_jC_k...)$ is either less than or greater than unity, then the solids are not in equilibrium with the gas. The quantities $a^c(A_iB_jC_k...)$ are called *pseudo-activities*.

For a given pressure and temperature, one can calculate the pseudo-activities by methods similar to those discussed above. These quantities define equilibrium $(a^c = 1)$, growth, $(a^c > 1)$ or decline $(a^c < 1)$ in the nucleation cluster. If growth occurs, then partial pressures in the gas must change. Nevertheless, these calculations can indicate the chemical nature of nucleation clusters most likely to form in the cooling gas.

In calculations[4] for AGB atmospheres of both O-rich and C-rich mixtures at assumed pressures of 10^{-10} bar, the materials likely to grow as nuclei can be identified. In the O-rich case, the material forming the first nuclei at the highest temperature is zirconium oxide, ZrO_2, at a temperature of approximately 1420 K, closely followed by corundum, Al_2O_3 (at a temperature of about 1380 K). Of course, there isn't much zirconium available (its cosmic abundance is less than carbon's by a factor of about a million, while that of aluminium is about one percent of carbon's abundance). So it is possible that the first significant nuclei to form under these physical conditions are aluminium oxide clusters, with a mere trace of zirconium oxide clusters. However, these nuclei cannot be the dust grains themselves as these materials are simply too rare to account for the amount of dust that is present. These nuclei are simply substrates on which more abundant materials may be deposited at a later stage to form grains.

In the C-rich example, zirconium carbide is the first material to form nuclei during the cooling process, closely followed by carbon. In perfect thermodynamic equilibrium, this carbon would be in the form of diamond. However, this form of solid carbon is difficult to achieve in low pressure environments, and a more likely form may be graphitic carbon, or soot (a mixture of solid carbons with both sp^2 and sp^3 bonding). However, it is worth noting that nanodiamonds have been found in some meteorites. Given the very low abundance of zirconium, the calculation suggests that the most important nucleation centres in the C-rich case are simply tiny carbon particles.

The growth of homogeneous nuclei can be addressed either by a kinetic theory, or – more usually in the astrophysical context – by a classical nucleation theory. The latter theory is used because of a lack of cluster data for use in rate equations, even though it cannot correctly account for the properties of small clusters. Both theories are beyond the scope of this book.

6.2.6 Growth of Dust Grains on Nuclei in AGB Stellar Envelopes

The growth mechanism is constrained by abundances of species that may contribute to growth. In the case of oxygen-rich outflows in which forsterite (Mg_2SiO_4) dust is present, growth must occur from the most abundant species relevant to forsterite. These species are SiO, Mg, and H_2O and a net reaction

$$SiO + 2Mg + 3H_2O \rightarrow Mg_2SiO_4(solid) + 3H_2$$

must proceed if growth is to occur. The precise details of this complex process are unclear.

Similarly, carbon dust grains must grow from the most abundant hydrocarbon, acetylene (C_2H_2), in the gas in a set of reactions summarized as

$$C_2H_2 \rightarrow 2C(\text{solid}) + H_2$$

where the solid is some form of soot. However, soot is composed largely of sheets of carbon with a graphene-like structure, in hexagons of carbon atoms. The bonding in these carbon sheets is aromatic, whereas the acetylene has a triple bond structure. A complex sequence of reactions is required to generate this overall reaction.

Grain growth may be opposed by evaporation or by chemical attack. However, if these processes are ignored, then a simple calculation shows that in the entire outflow, from the point at which dust growth begins to the point at which the dust grains emerge into the interstellar medium, the maximum radii of grains of any likely type are on the order of 0.1 µm. This value is consistent with the typical radii found for large interstellar grains, deduced from measurements of interstellar extinction (see Section 1.1.4). It is likely that a range of grain sizes will emerge from stellar envelopes. However, as we shall describe in Section 6.4.2, dust populations from cool stellar envelopes and other sources will – when ejected into the interstellar medium – be subject to a variety of modifying processes. The chemical structure may be modified, and their physical nature, including the size distribution, may be affected in situations of grain growth *via* grain–grain collisions or shattering or other erosion processes in shocks.

6.3   Dust Formation in Supernovae

6.3.1   Introduction to Supernovae

Supernovae (SNe) are stellar explosions that occur at the ends of the lives of massive stars (*i.e.*, with masses, say, of 8–30 $M_\odot$). A supernova (SN) is optically very bright indeed, and may at its peak be about as bright as the entire host galaxy of which the star is a member (see Figure 6.5a; this is an astounding situation – one star has become very briefly as bright as many billions of other stars). The rise in intensity is abrupt while the decay from the peak typically takes a few weeks or months. In this brief period, a supernova may emit as much energy (typically about 10^{51} erg) as the Sun will emit in its entire lifetime of ten billion years. The explosion drives much of the material of the

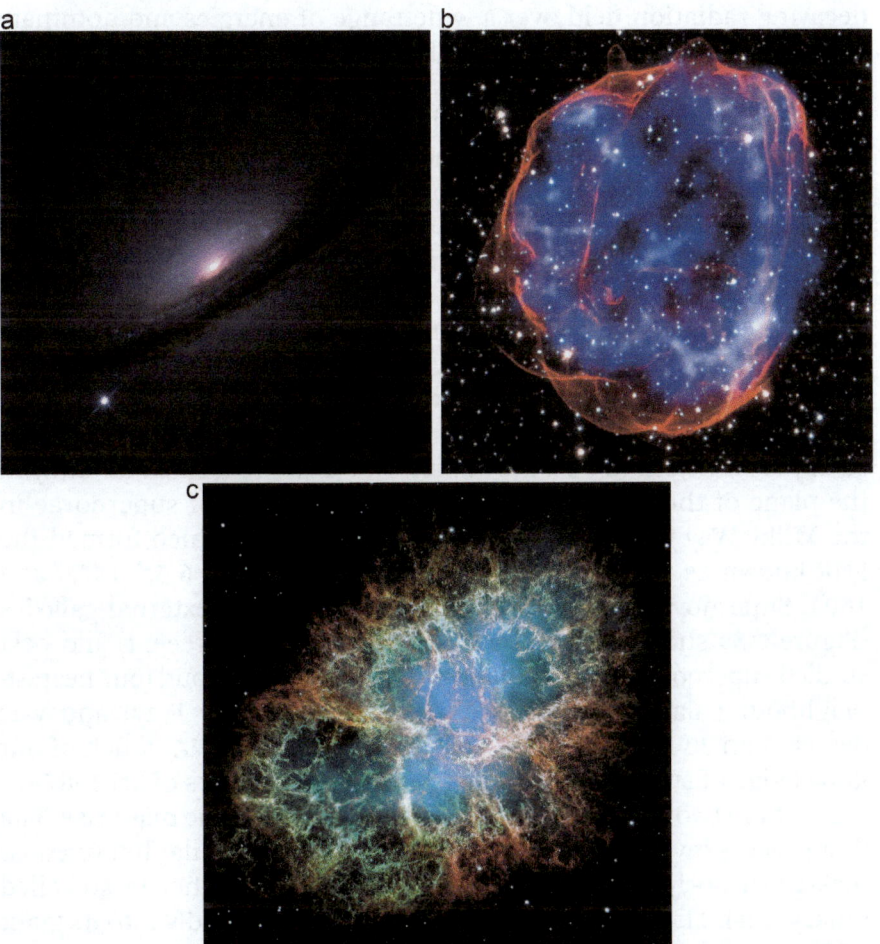

Figure 6.5 (a), (b), and (c) Images of a supernova and supernova remnants. (a) The supernova SN 1994D is of comparable brightness to its neighbouring galaxy NGC 4526 (credit: NASA/ESA, The Hubble Key Project Team and The High-z supernova Search Team). (b) An image of the supernova remnant SNR 0519690 created from X-ray emission from very hot gas (In blue, Chandra Observatory) and in visible light (Hubble Space Telescope) (credit for X-ray: NASA/CXC/Rutgers/J. Hughes; Optical: NASA/STScI). (c) The supernova remnant of a star that was observed to explode in 1054, known as the Crab Nebula (credit: NASA, ESA, J. Hester and A. Loll (Arizona State University)).

star into space, at astonishing speeds of tens of thousands kilometres per second. SNe are characterized by intense electromagnetic radiation, including γ-rays from a variety of radioactive nuclei (especially ^{56}Ni and ^{56}Co) from the stellar interior; these γ-rays generate a decaying radiation field over a wide range of energies and dominate the decaying light curve (the half lives of two isotopes of cobalt ^{56}Co and ^{57}Co are 77 and 272 days, respectively). The blast waves caused by the explosion sweep up a shell of interstellar matter, shocking it and raising its temperature to about a million K. This shell therefore becomes a very powerful emitter of radiation at all wavelengths from X-rays to radio, and is called a supernova remnant (SNR). Chemistry and dust formation must compete with this apparently very hostile environment. Images of two SNRs are shown in Figure 6.5b and c.

Supernovae are believed to occur in the Milky Way galaxy on average about two or three times per century. However, the last recorded supernova in the Milky Way was in 1604. Apparently, we are missing quite a few SNe; the explanation is that extinction in the plane of the galaxy is sufficient to obscure supernovae that are located far away in the plane of the galactic disc. However, detections of supernovae in the Milky Way were made in CE 185, 1006, 1054 (which formed the SNR known as the Crab Nebula illustrated in Figure 6.5c) 1572, and 1604. Supernovae are, however, readily detected in external galaxies (Figure 6.5a shows an example). The nearest, most recent, and best studied supernova occurred in the Large Magellanic Cloud (our nearest-neighbour galaxy, 50 Mpc distant from the Milky Way) and was detected on 23 February 1987; it is known as SN 1987A. Much of our knowledge of supernovae comes from detailed studies of SN 1987A.

There are two distinct processes in which supernovae may arise. The first process involves the evolution of two stars of similar but unequal masses in fairly close mutual orbit around each other (a so-called binary star). The more massive star evolves more rapidly into its giant phase, completes the conversion of hydrogen and helium to oxygen and carbon, but conversion to heavier elements doesn't occur and the star is on track to become a white dwarf. The less massive star then evolves into its giant phase and transfers large amounts of its material very quickly to the white dwarf because of the intense gravitational field generated by the white dwarf. The abrupt transfer of matter onto the white dwarf increases the pressure and temperature of the transferred materials so rapidly that nuclear reactions are triggered, synthesizing heavy elements throughout the star. These reactions inject enormous amounts of energy into the star, leading to the stellar explosion.

The second type occurs within a single massive star, rather than a binary. Nuclear reactions occur most rapidly at the centre of the star, and convert hydrogen nuclei to helium nuclei; helium nuclei are subsequently converted to carbon, neon, oxygen, silicon, and iron nuclei in exothermic reactions. However, iron has the most stable nucleus, and the reactions beyond iron are endothermic and the sequence ceases. The interior of the star then has a "shell" structure with iron nuclei at the centre and any remaining hydrogen at the edge and shells of carbon, neon, oxygen, and silicon nuclei in between (this is sometimes called the "onion-layer" model). When the star reaches this stage, no more nuclear reactions can occur, so the energy source supporting the star is suddenly removed and without thermal support (*i.e.*, pressure) the core of the star undergoes an abrupt collapse. Within a second, the star implodes, and material rebounds from the core in a huge explosion, leaving a neutron star or a black hole behind. This type of process is called a core-collapse supernova.

There are many divisions of these two types, based on classifications of supernovae spectra. The most abundant type of a core-collapse supernova has prominent hydrogen lines, indicating the presence of a significant amount of hydrogen in the stellar envelope. This envelope makes the outflowing ejecta massive and dense and therefore relatively slow moving. These characteristics favour dust formation, and core-collapse supernovae of this type (*i.e.*, with a significant hydrogen envelope) have been found to be the most significant sources of dust. A detailed discussion of supernovae and dust formation can be found in ref. 2.

6.3.2 Supernovae and Dust

It may seem surprising that supernovae can be significant sources of dust. The presence of very high speed shocks, high temperatures, and of intense short-wavelength radiation fields does not suggest that dust formation should be favourable in these locations. However, the observational evidence supporting the formation of substantial amounts of dust in many supernovae is very convincing. First, there is an observed excess in mid-infrared emission (wavelengths of about 10–100 µm) during the evolution of the supernova. This occurs precisely when the optical light curve declines sharply during the evolution. These two features are related: when dust formation occurs it tends to cause extinction of the optical emission, while absorption of the optical light by dust warms the dust so it emits more strongly in the infrared. Further, the peak of this excess infrared emission is observed to move to longer wavelengths, from about 30 µm at 3 y post explosion, to about 200 µm twenty years later; this peak wavelength change supports the idea that dust grains are growing to larger sizes over this period. In fact, there is clear observational evidence for dust formation from a few years post-explosion and dust grain growth for several decades.

More subtly, there is a systematic blue-shift of line emission profiles during the evolution. This is caused by the presence of dust extinguishing emission from the receding part of the ejecta (further away, from the point of view of the observer on Earth). The observer sees emission from that portion of the ejecta that approaches Earth (*i.e.*, blue-shifted emission) but not that from the more heavily extinguished receding (red-shifted) material, so the resulting profile appears to be blue-shifted.

Finally, there is evidence for the depletion of elements in the supernova, because the atoms and molecules containing these elements are incorporated into supernova dust. This is entirely similar to the depletion of elements in the diffuse interstellar medium that provides evidence for interstellar dust (and gives information about its composition), as discussed in Chapter 1, especially Table 1.4.

The observations of SN 1987A confirm that there is a growth in chemical complexity after outburst: first, some simple molecules appear; then dust appears and the mass of dust may grow substantially in time. Vibrational emission of CO (from level $v = 2$) was detected from about 100 days post explosion, and along with $v = 1$ emission, was seen for up to about 600 days. SiO rotational emission is seen after about 200 days. Emissions from H_2, HCO^+, and SO have also been detected. As the ejecta cooled, then rotational spectra were also seen. Evidently, there is a rudimentary chemistry present and evolving in the ejecta. At about the time that molecules were no longer detected (~600 days post explosion), dust emission made its appearance and was detected by its thermal emission. At 775 days post outburst, the mass of dust in SN 1987A was estimated from this emission to be about 10^{-4} $M_\odot$ and the dust temperature was then about 300 – 440 K. By day 1150 (about 3 years post outburst) the amount of dust was 5×10^{-4} $M_\odot$ at a temperature of about 150 K. Evidently, the growth of the dust mass seems to have accelerated in this period. By 2011, (about 24 years post-outburst) far-infrared measurements taken[5] using the Herschel Space Observatory determined the dust mass in SN 1987A to be ~0.3–0.7 $M_\odot$ at a temperature of 20–30 K. A few years later,[6] high resolution submillimetre waveband observations using the ground-based ALMA facility confirmed the dust mass to be ~0.5 $M_\odot$. Assuming that the interpretation of the earlier measurements was correctly made, the rate of growth of the dust mass is remarkably high. However, it is possible that much of the dust at very early times post-outburst was not detected because it was distributed in clumps and possibly hidden from observation. If so, substantial amounts of dust must have been formed at very early times post-outburst. Regardless of the correct interpretation, it is clear that amounts of dust on the order of a solar mass have been formed in SN 1987A (and by implication, many other SNe) within – at most – a few decades post-outburst. It is not yet clear how much of this dust will survive into much later times.

Mid-infrared spectra from supernovae can be compared with laboratory spectroscopic data and information about the composition of the supernova dust can be inferred. A wide range of silicates (including pyroxene, enstatite, ferrosilite, olivine, and fayalite) are inferred, along with quartz and various oxides of Al, Mg, Fe, and Ca. Carbon dust is also present, in the form of amorphous carbon and graphite. Carbon is also present in silicon carbide. Magnesium and iron sulfides have also been detected from observed infrared data.

6.3.3 Theoretical Approaches to Dust Formation in Supernovae

The observational data for SN 1987A and other SNe give clear evidence of chemistry and dust formation in the ejecta of supernovae. Is it possible to use a theoretical approach like that developed and used to describe chemistry in interstellar clouds (Chapter 5) and in circumstellar envelopes (Section 6.2, above)? This chemical kinetic approach is, as we have seen, potentially very powerful and has been used to understand the detailed chemical and physical processes involved in interstellar clouds. However, the physical conditions in the case of supernovae ejecta are very challenging, because they are far from steady state and from thermal equilibrium, and the ejecta move at very high speed, while the density and temperature of the ejecta decline rapidly. Nevertheless, the methods used in other interstellar situations can be applied to the ejecta, as long as the physical conditions can be described. Then, as in the other applications illustrated in this book, the evolving chemical composition in the ejecta can be explored using a large set of coupled chemical reactions. Chemistry leads to the formation of small molecules; these are assumed to form clusters which may act as nucleation centres on which molecules may accumulate; these clusters may coagulate to form large grains. Considerable work has followed these stages computationally (see ref. 2, which summarizes much of this work).

For a 15 solar mass progenitor star, by about 500–800 days post-outburst, the gas is believed to be mostly neutral, with number density $\sim 10^{10}$–10^9 cm^{-3} and temperature ~ 3000–1000 K (see Figure 6.6). The extent of mixing between shells, or between hydrogen from the progenitor star beyond the blast wave, is unclear but assumed to be

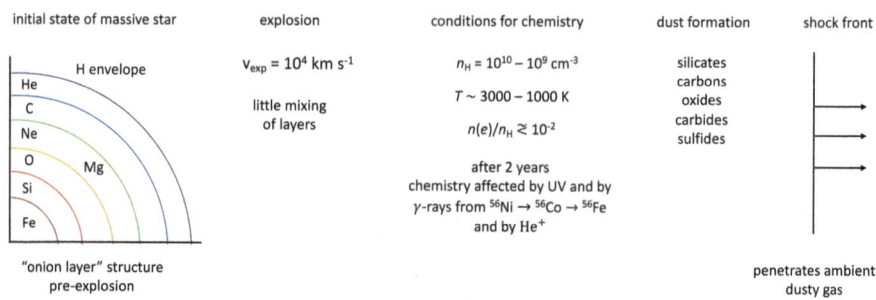

Figure 6.6 This sketch summarizes the conditions that lead to dust formation in the ejecta from a supernova.

poor. The theoretical model is assumed to be operating under these conditions.

The first stage, *i.e.*, exploring theoretically the chemical evolution, involves selecting a list of atoms, with the significant omission of hydrogen, and following a calculation of ions, molecules and eventually clusters formed under the relevant physical conditions. For about a dozen of the more abundant elements (omitting hydrogen), the scheme described in ref. 2 allows for the generation of a similar number of simple molecules appearing within a few years post-outburst. Carbon monoxide appears in these calculations by about 150 days post-outburst and becomes constant in abundance by about 700 days (consistent with observations of SN 1987A). The mass of CO does not decrease when grains are observed to begin to form, indicating that it is not a precursor to carbon dust formation. The chemical databases described in Chapter 5 can be used where possible. In fact, the restriction to H-poor chemistry greatly simplifies the reaction network.

Many of these species are available to create an extensive range of clusters. The formation of large carbon clusters is important, together with the formation of a wide range of clusters of metal oxides and silicates. The stage in which small clusters arise can be studied in a stochastic kinetically-driven approach. Important clusters at this stage include magnesium silicates, alumina, various silicon oxides, carbon structures including chains, rings, and fullerenes, and some other carbides, sulfides and oxides. Pure metal clusters are also predicted in these calculations. Clusters that attain a three dimensional structure tend to be more stable and are therefore favoured in these calculations.

These clusters then grow by coagulation and coalescence to form large grains (up to a few tens of nm). These processes are controlled by Brownian motion, turbulence, gravitational accumulation, and van der Waals forces (possibly affected by charge fluctuations on the grains). Models described in ref. 2 for a 15 $M_\odot$ star indicate that there is a sequence of dust formation of different types. At about 200 days post-outburst, iron sulfide grains appear. Forsterite grains follow at 300–400 days, and silicates appear at about 550 days. Carbon grains dominate the abundance of all the solid particles by about 1100 days. The total mass of dust by about 2000 days post-outburst in all forms is predicted in this model to be a few percent of a solar mass, and the grains are up to several tens of nm in size.

Although physically and chemically complex, this model almost certainly over-simplifies the true situation in a particular SN. However, the broad conclusions are impressive and generally in line with observations. It is clear from these theoretical investigations that conditions in core-collapse supernovae should be expected to generate a chemistry within a year post-outburst and that dust formation should occur within this material within a few years. The theoretically predicted dust mass is typically a few percent of a solar mass, and the chemical composition is consistent with results of observations of supernovae. Larger dust masses in the models may arise from clumpiness in the ejecta. However, the composition is related to the mass of the progenitor star. Larger stars generate large O shells and more O-rich dust.

When the high velocity ejecta encounter the interstellar gas surrounding the original progenitor star, they will experience a severe reverse shock. The effects of this reverse shock will significantly affect the chemistry and dust formed in the ejecta. It is important to note that these effects have not been described here.

We should also note that dust in the interstellar gas surrounding the progenitor star may also be affected by the sudden arrival of the supernova. The discussion given here has focused solely on new dust, formed in the ejecta. We have not considered the possible modification of pre-existing interstellar dust by the supernova.

6.4 Final Remarks on Dust Formation

6.4.1 Other Locations of Dust Formation

In this chapter, we have focused on dust formation in two main types of object, both of which represent very late stages in the evolution of stars: AGB stars and supernovae. These sources of dust are in fact the two that are predicted to produce the largest rates of dust injections to the Milky Way galaxy (see Table 6.4). However, there are other sources of dust that make significant contributions. Dust formation in novae may be important. The mechanisms by which dust is formed in novae are unique to novae, and are at least as complex as those found to be necessary in supernovae. Finally, massive stars in the late stages of evolution (WR stars) also contribute. While it seems unlikely that collisions of high speed winds from massive stars can be a source of dust, the fact that such an effect has been observed emphasises that dust can be formed under a wide variety of physical conditions.

6.4.2 What Happens to New Circumstellar Dust and New Supernova Dust in Interstellar Space?

There are many processes that affect newly-formed dust grains when they enter the interstellar medium. This region of space is pervaded by interstellar shock waves originating in non-local supernovae explosions. Passage of dust grains through shocks of sufficiently high velocity ($\sim$100 km s^{-1}) may cause grain–grain and gas–grain collisions. *Grain–grain* collisions may lead to *shattering*, redistributing grain mass from large grains to small grains and therefore changing the original grain size distribution to one that favours small grains to a greater extent. Shattering may also lead to *vaporization* of some grain material, and may remove some small grains entirely. Evidently, these two processes may have opposing results. The precise outcome of grain–grain collisions in high speed shocks depends on the nature of the dust and on the shock speed. *Gas–grain* collisions lead to *sputtering*, in which the grain surface is eroded so atomic and molecular material is ejected into the gas phase.

Of course, these destructive processes may be offset by processes causing growth of grains by the accretion of atoms and molecules from the gas phase on to the surfaces of dust grains. There is time

Table 6.4 The main contributors to dust formation in the Milky Way, with estimated injection rates (in units of $M_{\odot}$ pc^{-2} My^{-1}).

Source	Carbon dust	Silicate dust
AGB stars (C-rich)	3	
AGB stars (O-rich)		5
Wolf–Rayet	0.1	
Red supergiants		0.03
Novae	0.3	0.03
SN type Ia	0.3	2
SN type II	2	10

for such growth to occur in dense, dark interstellar molecular clouds. Grains may also grow by the coagulation of smaller units together into larger ones. The growth of dust grains into planetesimals in star and planet forming regions is an extreme example of this coagulation (see Chapter 9).

In low-density regions of the interstellar medium, the local radiation field may be capable of modifying the chemical nature of carbon dust. It is well-known that ultraviolet irradiation of H-rich sp^3 bonded carbon is capable of ejecting hydrogen atoms from the solid, converting it into a H-poor sp^2 material. This polymeric to aromatic conversion process also changes the optical properties and is therefore known as *photodarkening*. It can be reversed by *hydrogen insertion* if the H-poor sp^2 material passes through a high temperature region rich in atomic hydrogen. Evidently, at least part of interstellar dust responds to the local physical conditions.

Cosmic rays (see Section 1.1.3) may also affect the nature of interstellar dust, in particular the conversion of crystalline material to amorphous. It appears that these rays are effective in this role, in that nearly all the silicates in the interstellar medium are amorphous. Crystalline silicates have been detected in protoplanetary discs.

It is evident that grains produced in circumstellar regions (including SNe ejecta) are not necessarily the final material that might be found embedded in asteroids or in cometary dust. The interstellar medium contains a variety of materials at different stages of evolution. Some will be heavily processed during a long period in the interstellar medium and will be what we call interstellar dust. But other material will be pristine, largely unaffected since its formation in circumstellar regions. A good astrochemist should be able to read these materials for their histories, learning about the evolution of the Galaxy.

References

1. H.-P. Gail and E. Sedlmayr, *Physics and Chemistry of Circumstellar Dust Shells*, Cambridge University Press, 2014.
2. A. Sarangi, M. Matsuura and E. R. Micelotta, *Space Sci. Rev.*, 2018, **214**, 63.
3. E. Pigott, *Philos. Trans. R. Soc. London*, 1797, **87**, 133.
4. H.-P. Gail, in *Astrochemistry*, ed. Th. K. Henning, Springer-Verlag, Berlin, 2003.
5. M. Matsuura, *et al.*, *Science*, 2011, **333**, 1258.
6. M. Matsuura, *et al.*, *Astrophys. J.*, 2015, **800**, 50.

7 Surface Chemistry on Interstellar Dust Grains

"The most efficient mechanism of the formation of molecular hydrogen in the current universe is by association of hydrogen atoms on the surface of interstellar dust grains. The details of the processes of its formation and release from the grain are of great importance in the physical and chemical evolution of the space environments where it takes place." (J.-L. Lemaire, G. Vidali, S. Baouche, M. Chehrouri, H. Chaabouni, H. Mokrane)[1]

7.1 Active Roles for Dust Grains in Interstellar Clouds

In reading this book so far (up to and including Chapter 6), we may have gained the impression that dust grains are rather passive components in interstellar space: we have learned that they are formed as by-products of stellar evolution in stellar explosions and in stellar envelopes, and they are transported by those explosions and by stellar winds into interstellar space where their physical and chemical properties may be modified. In interstellar space, they passively absorb and scatter starlight, creating dark regions inside interstellar clouds from which starlight is mostly excluded. In another passive mode, dust grains lock up atoms of certain elements in solid form, so these atoms are not available for interstellar gas phase chemistry. We

Astrochemistry: Chemistry in Interstellar and Circumstellar Space
By David A. Williams and Cesare Cecchi-Pestellini
© David A. Williams and Cesare Cecchi-Pestellini 2023
Published by the Royal Society of Chemistry, www.rsc.org

shall see (in Chapter 9) that dust grains have important roles in the formation of planets; Earth – the planet we know best – is essentially an accumulation of processed interstellar dust; this is also a rather passive role for dust. It would be easy to conclude that the main role of dust in the chemistry of interstellar clouds is in passively shielding the interiors of clouds from the destructive effects of starlight, to allow the formation in gas phase chemistry of the remarkable variety of molecular species that is found in the denser regions of interstellar space (shown in the Appendix to Chapter 1).

However, dust grains have two very important *active* roles in interstellar chemistry. Firstly, the surfaces of dust grains are believed to be chemically active. Therefore, the reactions that generate the products of this surface chemistry should be considered along with the network of gas phase reactions that we discussed in Chapter 5. This surface chemistry is the topic of the present chapter. Secondly, under suitable physical conditions in the interstellar medium, dust grains accumulate icy coatings (often called icy *mantles*) composed initially of simple molecular species. These simple ices can be chemically processed through solid state reactions into more complex molecules, which may in appropriate circumstances be released from the solid state into the gas phase. These more complex molecules include many of the larger molecular species found in the Appendix to Chapter 1. The balance of formation of these larger species by gas phase chemistry, on the one hand, and by solid state chemistry, on the other, is currently of great interest to astrochemists. We shall discuss the chemistry of interstellar ices in the following chapter; the current chapter is devoted to surface chemistry on interstellar dust, a topic also dealt with at some length in ref. 2.

7.2 Do We Need Surface Chemistry on Interstellar Dust Grains?

The gas phase chemistry we have described in Chapters 3, 4 and 5 is elegant and to some extent fairly well-determined. The rate coefficients for many of the reactions can be studied in the laboratory or by quantum mechanical calculations. There is little doubt about the input and outputs involved in each gas phase reaction, and the computer models and databases discussed in Chapter 5 can easily cope with the complexity of having many thousands of such reactions competing with each other.

Why should astrochemists introduce surface reactions? These reactions are apparently not well-defined. They are taking place on surfaces that are poorly characterized on the atomic scale, either chemically or physically. The interaction of the reactants and their products with the surface is much more complex than interactions of reactants and products in the gas phase, and the ejection of the products from the surface may be a difficult problem that doesn't even exist in the gas phase case. So why invoke surface chemistry?

The immediate and simplest answer is: why not? We know that many surfaces act as catalysts in the laboratory and in industrial processes, so it is reasonable to consider whether the surfaces of dust grains may also have this capability in interstellar chemistry. A second and more compelling answer is that the gas phase schemes described so far in this book have at least one important omission: they do not explain the origin of interstellar molecular hydrogen in the Milky Way galaxy. There are probably a number of other failures of the gas phase schemes to account satisfactorily for the existence or the abundance of other molecular species in interstellar space. But molecular hydrogen is, as we have seen, the fundamental species that drives almost all of interstellar chemistry. So far, we have simply relied on observational information that shows that H_2 is a significant species in diffuse clouds, and is the totally dominant species in dark clouds and other dense interstellar regions. But we cannot claim to understand the main processes of astrochemistry if we don't know how H_2 is formed in the interstellar medium. After many decades of study, it is now accepted that molecular hydrogen in the Milky Way is formed almost entirely in surface reactions on interstellar dust grains. As we shall see below, this conclusion is supported by many detailed laboratory studies of systems that correspond quite well to those of dust in the interstellar medium. Strong support for this idea is also given by many fundamental theoretical (*i.e.*, quantum mechanical) studies of H_2 formation, although these reactions are envisaged to occur on somewhat idealized surfaces.

7.3   The Formation of H_2 in Interstellar Space

We have already mentioned several processes by which interstellar H_2 may be formed. The simplest idea is three-body association (see Section 3.2.11), but the number densities in interstellar clouds are many orders of magnitude lower than that required to enable this mechanism to supply H_2 at an adequate rate to compete with the loss by H_2 photodissociation caused by starlight (see Section 3.2.5). Three-body reactions are a viable mechanism under appropriate conditions, such as cool stellar atmospheres (see Chapter 6), but may be confidently excluded from consideration under the conditions found in typical interstellar clouds.

Another gas phase process is radiative electron attachment with hydrogen atoms:

$$H + e \rightarrow H^- + h\nu$$

The H^- is subject to loss in photodetachment by the interstellar infrared radiation field:

$$H^- + h\nu_{IR} \rightarrow H + e$$

Otherwise, a collision of the anion with another H-atom forms H_2:

$$H + H^- \rightarrow H_2 + e$$

This process is certainly viable and the relevant rate coefficients are accurately known. However, detailed studies show that this process is not efficient enough in the interstellar medium of the Milky Way to maintain the observed abundance of H_2, because dissociation of H_2 by starlight can be very fast (see Section 3.2.5). The problem with the H^- mechanism is that the initial attachment to form the anion is very slow while the photodetachment of H^- by the infrared component of the interstellar radiation field is very fast, so the intermediary, H^-, has a very low abundance.

A similar process involves the radiative association of H and H^+ to form H_2^+ which then undergoes a fast exchange reaction with H:

$$H_2^+ + H \rightarrow H_2 + H^+$$

However, similarly to H^-, the intermediate species H_2^+ is rapidly photodissociated by the visible and ultraviolet components of the interstellar radiation field, and so the overall H_2 production rate is not significant in the Milky Way. However, these two mechanisms may be important in dust-poor or dust-free regions, and especially in the Early Universe (see Section 5.7.1), before the elements of which dust grains are composed had been formed. These schemes are believed to be responsible for forming the very first hydrogen molecules in the dust-free Universe soon after the Big Bang (see Section 5.7.1).

The basic requirement for chemistry in the Milky Way is that the formation rate of H_2 must be large enough compared to the loss rate caused by photodissociation by starlight (see Section 3.2.5) to account for the H_2 abundances detected in diffuse clouds. Detailed studies of H_2 photodissociation, making proper allowance for the fact that the dissociations occurred in a spectral *line* process (rather than a continuum), were the first to show that the observed abundances measured along lines of sight through diffuse clouds to bright stars in the Milky Way could be accounted for by assuming H_2 formation in surface reactions (and loss in photodissociation by starlight) but not by other known formation processes.

In this picture, the roles of grains are to bring H atoms together on a surface (therefore, overcoming to some extent the problem of low interstellar gas number densities), to absorb some of the energy released when the H_2 molecule is formed (so the newly-formed molecule remains bound), and – possibly – to use part of the energy released in molecule formation to eject the newly-formed H_2 molecule from the surface into the gas phase. The detailed study of the balance of H_2 production and destruction in diffuse clouds then requires that most of the H atoms striking a dust grain would leave the surface of the grain as part of a H_2 molecule.

The detailed studies of all known H_2 formation mechanisms showed that the balance of H_2 formation and destruction in diffuse clouds could not be met in the Milky Way by any H_2 formation process except surface reactions on dust grains. However, the formation of H_2 on grain surfaces is required to be an efficient process, occurring for almost every collision between a H atom and a grain. Is that plausible?

We can estimate the rate of formation of H_2 as described in the previous paragraph in a simple calculation. We write

$$dn(H_2)/dt = n(H)\, v_H <\pi a^2\, n_d>\, \varepsilon$$

where $n(H)\, v_H$ is the flux of H atoms per unit area arriving at the surfaces of grains, and $<\pi a^2\, n_d>$ measures the total cross sectional area of all grains of all radii, per unit volume of space (assuming that the grains are spherical, which is – of course – not necessarily the case). Finally, ε is a parameter that allows for the overall efficiency of the formation process, including sticking, mobility, reaction and ejection of the newly-formed H_2 from the surface (these factors are discussed in the following Section 7.5).

For illustrative purposes, let's assume that the grains causing visual extinction are those that contribute to H_2 formation. These grains have sizes comparable to the wavelength of visible light, and from extinction measurements we infer that there are about $n_g \sim 10^{-12}\, n_H$ of these grains per unit volume, where $n_H = n(H) + 2\, n(H_2)$ is the total number density of hydrogen atoms in both atomic and molecular forms. With some rearrangement, we can then write the approximate H_2 formation rate in units of $cm^{-3}\, s^{-1}$ as

$$dn(H_2)/dt \sim 3 \times 10^{-17}\, n(H)\, n_H\, (T/100\ K)^{1/2}\, \varepsilon$$

where we have displayed the temperature dependence separately from the velocity, v_H, and the number densities $n(H)$ and n_H are both in units of cm^{-3}. Since we know that the process must be very efficient if H_2 formation in diffuse clouds is to compete with photodissociation by starlight, the efficiency parameter must be large, less than but on the order of unity. In fact, if ε is about ½, then this formation rate is adequate to account for the formation of molecular hydrogen in diffuse clouds.

We noted in Chapter 1 (Section 1.1.4) that there is a size distribution of grain sizes, with very many more small grains than large. Therefore, the expression $<\pi a^2 n_g>$ must take this into account. In fact, the amount of cross section per unit volume of space available for H_2 formation is dominated by the surfaces of the smallest grains in this distribution, and the total may be several times larger than the value used above. The number and radii of the smallest grains are not well determined.

In dark clouds, molecular hydrogen is destroyed by cosmic ray ionization, producing H_2^+ and $H + H$ in about equal channels. The reaction between H_2^+ and H_2 releases one more H atom. Exchange reactions between hydrides and other non-H atoms release more atomic hydrogen. These H-atoms are gradually returned to hydrogen molecules by surface reactions, so the abundance of molecular hydrogen in dark clouds is always maintained.

7.4 Possible Mechanisms for the Formation of H₂ in Surface Reactions

The studies of molecular hydrogen in diffuse clouds re-invigorated investigation of the simple idea that surface reactions on dust grains could – in principle – form enough H_2, but left the details of the process completely undefined. In fact, on closer investigation, these details are fairly complex. First, a H atom must arrive at a grain surface and be retained there at least until a second H atom arrives. This constrains the grain temperature, because desorption is faster from a warmer grain. Second, the two atoms must be able to find each other before one of them is desorbed, so one of them (at least) must be mobile on the surface and must therefore be physisorbed (bonded by van der Waals forces to the grain surface in a potential well of depth a few ~ meV) rather than chemisorbed (bond energies ~ eV). If at least one atom is physisorbed, it may roam over the surface with high mobility (by quantum mechanical tunnelling through small energy barriers) until it locates and reacts with the other atom. This is known as the Langmuir–Hinshelwood mechanism (see Figure 7.1a) and is the case for low gas densities and low surface coverage. The alternative is that the second atom collides with the surface sufficiently close to the first so that it reacts immediately without having to search the surface. This is called the Eley-Rideal mechanism (see Figure 7.1b). Obviously, the Eley-Rideal mechanism is more likely if the surface coverage is high.

In either case, it is assumed that when the two atoms are held close together by the surface, a reaction occurs and the molecule is formed. The molecule may be either desorbed promptly with some kinetic and rovibrational energy, depositing some energy into the surface, or it may be retained on the surface, transferring energy released during the formation process into the surface and becoming thermalized at the temperature of the grain. The retained molecule may

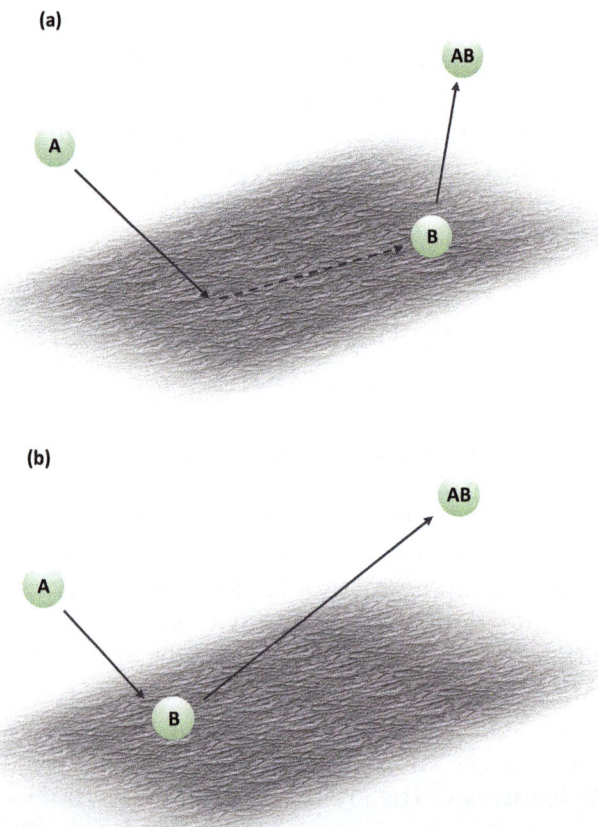

Figure 7.1 (a) Langmuir–Hinshelwood and (b) Eley-Rideal mechanisms.

then be desorbed at a later stage, perhaps during a transient heating event. Figure 7.1 shows the prompt desorption for both Langmuir–Hinshelwood and Eley-Rideal cases, but of course the products in each case may also be retained on the surface and desorbed later in a separate process.

The Langmuir–Hinshelwood and Eley-Rideal mechanisms are simple idealizations of what may be a more complicated situation. For example, given that the gas temperature in diffuse clouds (~100 K) is much greater than the temperature of dust grains in those clouds (~10 K, estimated from thermal radiation measurements, and from thermal balance calculations), H atoms are "hot" compared to the dust, and several collisions, each with declining energy, may be necessary for the incident atom to be thermalized. The "hot" atom may encounter another atom bound to the surface during this thermalization process. This "hot atom" mechanism is conceptually somewhere between the Langmuir–Hinshelwood and Eley-Rideal idealizations.

7.5 Detailed Studies of H₂ Formation

7.5.1 Both Theoretical and Experimental Studies are Possible and Required

There has been a huge amount of both theoretical and experimental work on the formation of H_2 on surfaces that may represent grains. These approaches complement each other. Theory enables the nature of the surface to be defined on the atomic scale, *i.e.*, is the surface a perfect crystal, or is it deformed in some way? Is the material locally crystalline or amorphous? Theoretical studies can probe the effectiveness of H_2 formation on the detailed atomic structure of the surface. On the other hand, experimental studies must use bulk materials that may represent the global average nature of interstellar dust but not the local properties. These studies must involve many kinds of surface sites so these experiments must give some kind of average of the effectiveness of H_2 formation on these surfaces.

7.5.2 What Can We Learn from Theoretical Studies?

For ease of calculation, much of the theoretical work has adopted a well-defined graphene-type sp^2 surface as the model for the surface of the interstellar grain. The calculated H-atom physisorption and chemisorption bond energies are computed to be close to the measured values (39.2 meV and 0.67 eV, respectively), suggesting that this theoretical approach is reliable. Many studies have suggested that chemisorption on a graphene-type surface distorts the plane by raising the carbon atom to which bonding occurs slightly out of the graphene plane. This distortion creates a barrier of about 0.2 eV in the interaction of the H atoms with the surface, apparently inhibiting chemisorption under cool interstellar conditions. However, a more detailed theoretical study[3] (and some experimental work) showed that this barrier is reduced significantly when there are other chemisorbed H atoms near to the site. Therefore, it seems that chemisorption is able to occur on a model carbon surface for interstellar dust. However, not all the possible paths for an incident H atom to interact with a chemisorbed H-atom are permissible at low temperatures. Some paths are inhibited by an energy barrier, while other paths are accessible so a reaction will occur.

Computed mobilities for physisorbed atoms are high, but the sticking probability of these atoms is rather low. In these calculations, both Langmuir–Hinshelwood and Eley–Rideal models appear viable, with the Eley-Rideal process more effective than the Langmuir–Hinshelwood one in producing H_2.

Theoretical studies generally indicate that the reaction probabilities are very high, with the energy released on H_2 formation (~4.5 eV in total) shared between a very large amount of rovibrational excitation in the newly-formed molecule, with a significant amount of energy transferred into the lattice, and some kinetic energy.[4] These results suggest that if H_2 is desorbed on formation then the newly-formed molecule may be expected to be rovibrationally and kinetically excited. However, it is also possible that some newly-formed H_2 molecules remain trapped and thermalized at the surface because of the nature of the particular site. If so, their energies of formation must be transmitted to the surface and be absorbed into (and heating) the grain.

7.5.3   What Can We Learn from Experimental Studies?

In spite of the exceptional difficulties associated with these types of experiment, there have been many laboratory studies of H_2 formation by surface chemistry carried out over the last two decades, starting with ground-breaking work[5] by Valerio Pironello, Gianfranco Vidali, and colleagues. The technical problems arise from working at ultra-high vacuum (10^{-11}–10^{-10} Torr), at low gas temperatures (~100 K) and at even lower surface temperatures (~10 K), and in devising a suitable signal to detect small quantities of newly-formed H_2 molecules. These difficulties are indeed formidable; however, they have all been successfully overcome. The overall conclusion from experimental studies of H_2 formation in surface reactions is that these reactions are successful in forming H_2 molecules under conditions that represent those found in interstellar clouds as closely as possible in a laboratory situation.

A commonly adopted approach is to study the formation of HD, rather than H_2. This requires the use of two atomic beamlines, one for H and one for D, the atoms being produced by radiofrequency dissociation of H_2 and D_2. These beams are cooled appropriately to the desired gas temperature and impinge on the selected cold surface representing the dust grain, also cooled to the desired (much lower) dust temperature. Product molecules then arise from reactions between H and D at the surface to form HD; this product molecule is quite distinct from any residual H_2 and D_2 in the reaction chamber from the beamlines. Quadrupole Mass Spectrometry (QMS) is routinely used in these experiments to determine abundances in the H and D beams and in the HD formation region. Resonance Enhanced Multi Photon Ionization (REMPI) is often used to interrogate the rovibrational state of the newly-formed molecule.

Some experiments measure the amount of the product molecule retained at the surface by following a Temperature Programmed Desorption (TPD) experiment after the irradiation of the surface is complete. In a TPD experiment, the sample surface is slowly warmed and the released product molecules examined by QMS. These experiments show that a large amount of product molecules can be formed on surfaces with temperatures in the range ~10–15 K and these product molecules can be retained on the surface until released by warming in the TPD experiments.

Other experiments have examined the formation of molecules released promptly from the surface by studying (using REMPI) the rovibrational excitation in the newly-formed molecules. The population in a particular excited level can be used as a measure of the formation rate of molecules in the case of prompt release. In some experiments using a surface of Highly Oriented Pyrolytic Graphite (HOPG) at a surface temperature of 15 K, excitations of the product molecule in a variety of levels up to $v'' = 7$ and $J'' = 4$ have been found.[6] The rovibrational level $v'' = 4\ J'' = 2$ is well-populated in the H_2 formation process, and has been used as a signature of molecular hydrogen formation. These studies also indicate that a large amount of energy is deposited into the surface at the reaction site, and that the molecule departs from the surface with up to about ~ 1 eV of kinetic energy. The energy deposited into the surface will cause local heating and may contribute to desorption of other nearby adsorbed species. These experimental studies of the energy budget in surface reactions confirm the theoretical predictions described in Section 7.5.2.

Experiments like these have demonstrated that there are two types of process occurring on analogues of dust grain surfaces. In one type, the product molecules are retained on the surface until they are released in TPD warming. In the second type, product molecules are released promptly, with considerable rovibrational and kinetic energy, in a process in which large amounts of energy are given to the surface. A comprehensive investigation[1] has determined that both types of process may occur simultaneously in the same experiment. The first type, in which the product molecules are retained and only released under TPD, is found to occur in a narrow surface temperature range (10–15 K). The second type, involving prompt release and high rovibrational energy excitation, appears to occur over a very much wider surface temperature range (10–70 K). This result confirms that there are (at least) two types of surface site at which reactions may occur. The first type (retained products) may be essentially the Langmuir–Hinshelwood mechanism, while the second (prompt desorption) may be the Eley-Rideal process in operation. Both mechanisms can contribute to H_2 formation in the interstellar medium.

7.6 Can Molecules Other than H_2 be Formed in Surface Reactions on Dust Grains?

Suppose that we want to account for observed molecular abundances in a particular astronomical object. We adopt the physical description of the region proposed by the astronomers; say, a diffuse cloud of a particular density and radiation field intensity. We follow the chemical prescriptions listed in Chapter 3 and we utilize them in gas phase chemical networks as described in Chapters 4 and 5, adopting the best available data on reaction rate coefficients from the latest databases (such as UDfA and KIDA). Let's imagine that we find that the results of our computations predict some molecular abundances that differ greatly from those observed (usually, a failure of this kind means that our computational model abundances are much smaller than those deduced from the observational data). What then?

It may be that there are problems with the interpretation of the observations, so that the physical conditions we have adopted in our computational model are incorrect. In that case, it is easy to repeat the calculations with modified parameters. We may be able to improve the match of the computations with observations and so revise our understanding of the physical conditions in that region of interstellar space.

Alternatively, it may be that some of the gas phase rate coefficient data are incorrect. If so, we may be able to trace errors of this kind by the effect they have throughout the network. Then the incorrect rate coefficients must be revised by experiment or theory.

Or it may be that the chemical network is incomplete and that surface reactions should be included in the network. For example, if we excluded from our chemical network equations to describe the formation of molecular hydrogen on grain surfaces, that network would be incomplete and such a model would predict that the abundance of molecular hydrogen would be very low and that the abundance of species depending on the presence of H_2 would be negligible. Could it happen that other surface reactions occur, forming other interstellar molecules, and that these reactions should also be included in the network?

Computational models of diffuse cloud chemistry based on the ideas of Chapter 3 incorporated in the chemical networks of Chapter 4, and using the latest estimates of all the reaction rate coefficients, predict results that are generally in reasonable agreement with molecular abundances determined from observational data for simple oxygen chemistry and carbon chemistry. However, the abundances predicted by the computational modelling for the nitrogen hydrides NH, NH_2, and NH_3 seem to be too low compared to the values obtained from the observational data.

It is quite possible that surface reactions contribute molecules other than H_2 to the chemistry. There are several experimental studies showing that sequential hydrogenation of nitrogen atoms in reactions on surfaces occurs efficiently in a barrier-free process, suggesting that such processes should occur even at the low temperatures of interstellar dust in diffuse clouds.[7] Consequently, a modelling study has been made[8] in which O and N atoms strike grain surfaces and leave as hydrides, for example,

$$N + g\text{-}H \rightarrow NH + g$$

(where g is a dust grain; similar reactions form NH_2 and NH_3; and other reactions form analogous oxygen species). These reactions are assumed to have the same efficiency as H_2 production on surfaces, but with allowance for the lower abundances and lower thermal speeds of O and N compared to H atoms. This study indicates that the oxygen-bearing molecular abundances produced by the gas phase network (which agree fairly well with the observational data) are hardly changed by the inclusion of surface reactions. However, the discrepancy between the results from modelling and from observational results for nitrogen hydrides is removed when surface nitrogen chemistry is included. According to this study, surface reactions can contribute very effectively to the formation of nitrogen hydrides, and more effectively than gas phase schemes. Indeed, the study also suggests that nitrogen hydrides are injected from the surface reactions in equal amounts of NH, NH_2, and NH_3. By contrast, oxygen hydride chemistry in diffuse clouds is determined almost entirely by the gas phase reactions described in Chapter 3.

References

1. J.-L. Lemaire, G. Vidali, S. Baouche, M. Chehrouri, H. Chaabouni and H. Mokrane, *Astrophys. J.*, 2010, **725**, L156.
2. D. A. Williams and C. Cecchi-Pestellini, *The Chemistry of Cosmic Dust*, Royal Society of Chemistry, 2016.
3. B. J. Irving, A. J. H. M. Meijer and D. Morgan, *Phys. Scr.*, 2011, **415**, 028108.
4. B. Kerkeni and D. C. Clary, *Chem. Phys.*, 2007, **338**, 1.
5. V. Pironello, C. Liu, L. Chen and G. Vidali, *Astrophys. J.*, 1997, **475**, L69.
6. E. R. Latimer, F. Islam and S. D. Price, *Chem. Phys. Lett.*, 2008, **455**, 174.
7. H. Linnartz, S. Ioppolo and G. Sedoveev, *Int. Rev. Phys. Chem.*, 2015, **34**, 205.
8. Z. Awad, S. Viti and D. A. Williams, *Astrophys. J.*, 2016, **826**, 207.

8 Interstellar Ices and Solid-state Chemistry as a Route to Molecular Complexity

"More than seventy complex organic molecules were detected towards various interstellar and circumstellar regions. With the exception of methanol, these complex organic molecules were only detected in the gas phase. But their sole existence challenges the scheme of gas-phase reactions and suggests a formation in ice-covered dust grains or in the ice–gas interface." (G. M. Muñoz Caro, A. Ciaravella, A. Jiménez-Escobar, C. Cecchi-Pestellini, C. González-Díaz and Y.-J. Chen)[1]

8.1 Ice Mantles on Dust Grains in Interstellar Clouds: A Route to Chemical Complexity

8.1.1 Observational Information About Interstellar Ice

Water ice was first detected in interstellar clouds in 1973 in observations by Gillett & Forrest.[2] They found absorption in the pure O–H stretching mode of H_2O, generating a broad spectral feature centred near 3 μm. The rotational structure that would appear in absorption from gaseous H_2O molecules is suppressed in solid ice since the molecules are not free to rotate; this rotational structure is missing in

Astrochemistry: Chemistry in Interstellar and Circumstellar Space
By David A. Williams and Cesare Cecchi-Pestellini
© David A. Williams and Cesare Cecchi-Pestellini 2023
Published by the Royal Society of Chemistry, www.rsc.org

the observational spectrum, so there is no doubt that the detected feature arises in solid material. The absorptions near 3 μm are found when radiation emitted from a background source (such as an ionized region or a distant star) passes through a foreground interstellar cloud containing water ice, on its way towards detection on Earth. A number of other modes of solid H_2O, from 4.5 to 63 μm, exist and have been detected in astronomical surveys.

While the ice is found on many lines of sight, it is detected only in denser, darker interstellar regions and is not present in regions in which the interstellar extinction is low. Figure 8.1[3] shows observational results for several lines of sight on different paths through the rather clumpy Taurus Molecular Cloud. The optical depth of the 3 μm

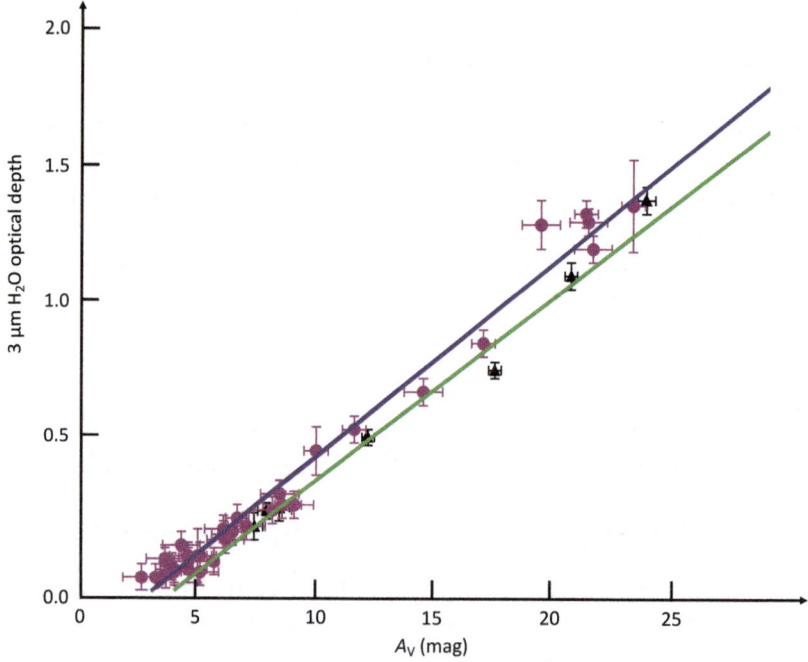

Figure 8.1 Plot of optical depth in 3 μm H_2O absorption against visual extinction along various lines of sight through two clumpy clouds (Taurus, with purple symbols, and IC 5146 with black symbols) towards background stars; these lines may pass through dense regions of the clouds (with high extinction) or more tenuous regions (with reduced extinction). The observational results show that high extinction paths show a strong 3 μm absorption, while low extinction paths show a weak or zero 3 μm absorption. There is a threshold extinction of about 3 visual magnitudes, below which no 3 μm absorption occurs. The two clouds show very similar behaviour. The figure uses data from ref. 3.

H_2O absorption is plotted against the visual extinction A_V, in magnitudes, for each line of sight.

This figure shows that lines of sight through regions of the cloud with high visual extinction have relatively strong absorption at 3 μm, and *vice versa*, and that there is a reasonably near-linear dependence between the optical depth and the corresponding visual extinction. The figure also implies that there is a minimum value of A_V for detection of water ice; this critical value, A_{Vcrit}, is about 3 magnitudes in the case of the Taurus Molecular Cloud, but may be quite different in other molecular clouds. The observations suggest that in regions with A_V above this critical value, A_{Vcrit}, ice is deposited onto the surfaces of interstellar grains.

In fact, the implications of the observations go further. Since the abundances of H_2O molecules in the gas are relatively low in interstellar clouds (though not in shocked regions), there are not enough H_2O molecules in the gas which might collide and stick to grain surfaces to create the detected ice layers within the lifetime of the cloud. However, the free oxygen atom abundance is high, so the existence of an ice layer, or mantle, on grain surfaces requires that O atoms colliding with a grain undergo a surface reaction to form H_2O molecules, and that at least some of these product H_2O molecules are retained on the grain surface in the form of water ice. Therefore, the existence of interstellar ice is confirmation that the surfaces of grains are chemically active. The surface reactions by which H_2O is formed are described in Section 8.2. Evidently, in regions in which the visual extinction, A_V, is less than the critical value implied in Figure 8.1, A_{Vcrit}, a process is operating that either inhibits the formation of H_2O on the surface or ejects a newly-formed H_2O molecule promptly into the gas where it is swiftly destroyed by photodissociation by starlight.

A detailed study[4] of the comprehensive chemistry in a dynamically evolving interstellar cloud of typical cloud density shows that the characteristic shape of the ice abundance curve of the kind seen in Figure 8.1 arises naturally. The slopes of the linear portions of these curves arise from balancing the freeze-out of oxygen atoms with photodesorption of H_2O molecules. The critical extinction for the onset of ice deposition is shown to depend on the local density and local radiation field. Most of the ice in a cloud is deposited on very small grains (since they are much more abundant than large grains). Hence the dust grain size distribution and the consequent ultraviolet extinction within the cloud are greatly modified in a cloud in which ice is deposited on grain surfaces.

Water ice mantles are observed to be very far from being chemically pure. High resolution infrared spectroscopy reveals a variety of species in ices observed along various interstellar lines of sight, see Figure 8.2.[5] The figure shows spectra observed towards several lines of sight through dense cold interstellar clouds, and the features are identified by comparison with laboratory data obtained for various ice mixtures. Note that the number of species detected and present in interstellar ice is very small compared to the number detected in interstellar and circumstellar gas (see the Appendix to Chapter 1).

The figure shows that there is some variation in composition from one line of sight to another, suggesting that the chemical composition of the ice responds to the nature of its environment. We can explore these variations by separating observational data into various groups depending on the nature of the environment in which the cloud showing the molecular features is located, see Table 8.1.[6] The groups are quiescent clouds (*i.e.*, unaffected by star formation), clouds in regions of low-mass star formation (*i.e.*, where electromagnetic radiation is somewhat enhanced above the interstellar medium), and clouds in regions of high-mass star formation (where electromagnetic and particle radiations may be substantially enhanced).

It is clear that the different environments significantly influence the chemical composition, reflecting the greater intensity of optical and X-ray emissions and energetic particles in star-forming regions than

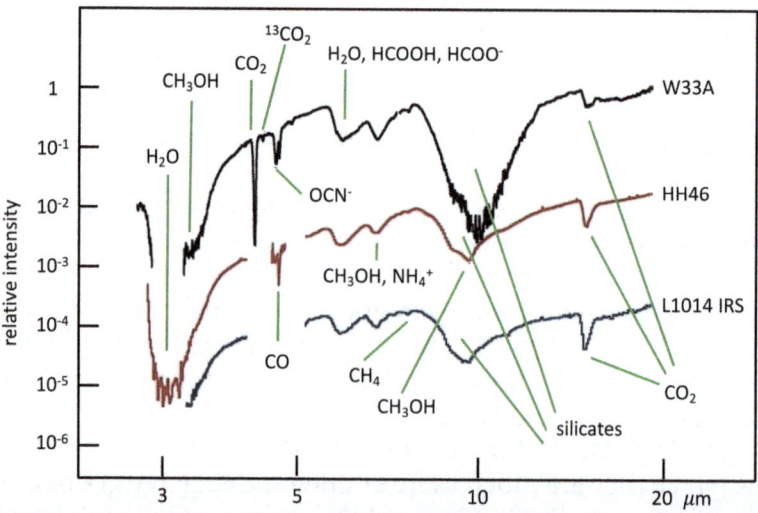

Figure 8.2 High resolution infrared spectra of several stars in the Milky Way, along lines of sight passing through different dense cold clouds (data taken from ref. 5). Absorptions caused by various solid state molecular species are present in these spectra.

Table 8.1 The relative composition of interstellar ice in different environments (from Tielens 2021[6]).

Species	Quiescent cloud	Low mass protostar	High mass protostar
H_2O	100	100	100
CO(total)	25	5	13
CO (H_2O ice)	3	10	6
CO(pure)	22	4	3
CO_2(total)	21	19	13
CO_2 (H_2O ice)	18	10	9
CO_2 (CO mix)	3	10	2
CO_2–CH_3OH complex	—	—	1
CO_2 (pure)	0	0	1
CH_4	<3	<1	1.5
CH_3OH	10	30	18
H_2CO	0	0	6
HCOOH	<1	<1	7
OCS	<0.2	<0.1	0.2
NH_3	<8	<11	15
OCN^-	<0.5	<0.2	0.1

in quiescent clouds. We may therefore regard the nature of ice in quiescent clouds as the ice composition as it is initially deposited, while the nature of ice in low mass and high mass star-forming regions represents ice that is increasingly processed into revised chemical compositions. This is important evidence that these mixed ices are not inert materials, but instead can be wonderfully reactive. We shall see in Section 8.3 that under suitable conditions these interstellar ices may act as a chemical feedstock, potentially capable of producing a range of about a hundred or more complex molecular species, almost all of which are organic; many of these molecular species are listed in the Appendix to Chapter 1, showing the detected interstellar and circumstellar species. For the moment, however, we shall discuss how the variety of species listed in Table 8.1 is believed to arise. This chemistry is largely driven by the arrival and retention of CO molecules at the surfaces of dust grains, and is described in the following Section 8.2.

8.2 Ice Chemistry in Quiescent Interstellar Clouds and in Clouds in Regions of Low and High Mass Star Formation

Laboratory studies[7–9] of many surface reactions that may form surface H_2O have identified three mechanisms that operate without energy barriers and may be expected to occur readily on the surfaces of dust

grains in cold interstellar clouds at temperatures ~10 K. These are
(i) successive H additions to surface O-atoms to form surface OH
and then surface H_2O; (ii) a complex sequence of reactions involving
the hydrogenation of surface molecular oxygen, O_2, to form surface
OH and H_2O; and (iii) reactions hydrogenating surface ozone, O_3, to
form surface OH and then H_2O. It may seem surprising that O_2 and O_3
may be involved in interstellar surface chemistry as these molecules
are rare in the gas phase. However, in the gas phase in dense clouds,
we have seen that carbon monoxide takes up nearly all the carbon,
leaving a significant excess of free oxygen, while hydrogen is almost
entirely in H_2 and atomic H is a minor constituent; the result is that
the abundances of atomic hydrogen and atomic oxygen arriving at
the surface of a dust grain may be comparable. Therefore, an O atom
bound to the surface of a dust grain in a dark cloud is quite likely to
be hit by another O atom to form a surface O_2, and similarly a surface
O_2 molecule may be converted to O_3. The three barrier-free reaction
schemes all involve reactions with H atoms and are shown here:

(i) $O \rightarrow OH \rightarrow H_2O$
(ii) $O_2 \rightarrow HO_2 \rightarrow H_2O_2 \rightarrow OH, H_2O$
(iii) $O_3 \rightarrow OH \rightarrow H_2O$

where the arrows signify reaction with hydrogen. The reactions at low
temperatures occur by the Langmuir–Hinshelwood mechanism (see
Chapter 7, Section 7.4) and at each stage the products are stabilized
by rapid energy transfer to the grain.

In quiescent clouds, CO-ice appears mainly as pure CO-ice, with only
a small amount of CO–H_2O mixed ice. However, CO-ice is fairly heavily
converted to CO_2 ice and to ices of more complex species. The formation
of CO_2 ice in mixed ices at low temperature by the addition of O atoms
is found to be slow in experiments, because the O atoms preferentially
form O_2 and O_3. A more likely surface route is the reaction between OH
(from the photodissociation of H_2O) and CO. In fact, this is a reaction
that forms CO_2 very efficiently in gas phase interstellar chemistry. The
similar surface reaction will be aided by the fact that newly-formed OH
will be kinetically excited and can locate its CO partner rapidly.

All three types of interstellar cloud in which cold dust is identified
show substantial amounts of methanol ice. While the sequential addition of H atoms to CO

$$CO \rightarrow HCO \rightarrow H_2CO \rightarrow H_3CO \rightarrow CH_3OH$$

might seem to be the obvious route, laboratory studies and theoretical investigations[6] show that two of the steps in this sequence

are impeded by appreciable energy barriers. This simple sequence to methanol may not be able to operate at the low temperatures of interstellar clouds.

In spite of these theoretical impediments, laboratory experiments[10] show that – *when these ices are irradiated by H_2 Lyman and Werner band emission* – methanol is indeed formed by solid-state chemistry in a H_2O/CO ice mixture (4 : 1) or in a pure ice layer with a monolayer of CO on the surface. The irradiation is fairly similar to the mean interstellar ultraviolet radiation field generated by cosmic rays inside an interstellar dark cloud (see Chapter 3, Section 3.2.6). The duration of the irradiation in these experiments was chosen to match the radiation dose that would be received inside an interstellar cloud in a period of about 10^5 years. Other experiments without irradiation but with a H-atom flux were also performed. In all cases, the CO signal declined (in part because CO is efficiently photo-ejected from the ice), while those of CO_2, H_2CO, HCOOH and CH_3OH grew. Evidently, under appropriate laboratory conditions CO can be converted to simple hydrocarbons such as formaldehyde and methanol. These conversion reactions can readily account for the presence of small hydrocarbons in the different types of ice composition shown in Table 8.1.

Methane and ammonia ices have not been detected in interstellar ice in quiescent clouds, nor in clouds associated with low mass star-forming regions. They are, however, detected in clouds in regions of high mass formation, which suggests that strong radiation fields may promote the formation of these molecules. Methane in the ices is readily formed by sequential low temperature surface reactions of atomic carbon with H atoms. However, since most of the carbon in interstellar clouds is tied up in gaseous CO, the abundance of free carbon in the gas, and consequently incorporated in ices on grains, is relatively low. Nitrogen atoms may behave similarly. However, ammonia is present in interstellar clouds, and will naturally be incorporated into ices on grain surfaces. The detection of ammonia ice only in regions of high-mass star formation (which are of higher density) suggests that the greater gas densities in these regions are responsible.

8.3 Complex Molecules from Simple Interstellar Ices

8.3.1 Some Theoretical Considerations

We have seen in this chapter so far that interstellar ices are observed to be deposited on the surfaces of dust grains in interstellar clouds where the visual extinction is above some critical value, A_{Vcrit}. These ices are initially deposited as pure water ice with a later deposition of CO ice. Then, assisted by irradiation by local radiation fields, reactions occurring in these initial ices form the variety of species listed in Table 8.1. Therefore, a significant amount of chemical complexity has been generated in the solid state by reactions in the initial H_2O and CO ice. In particular, organic species such as methanol, CH_3OH, are widely detected in interstellar ices, and methane (CH_4), formic acid (HCOOH), formaldehyde (H_2CO), ammonia (NH_3), carbonyl sulfide (OCS), and the cyanate ion (OCN^-) are found in interstellar ices and gas associated with high-mass star forming regions. Given that these species are formed in solid-state reactions, is it possible that these interstellar ices may in some circumstances initiate a further sequence of reactions that generate even more species with even greater chemical complexity?

Garrod et al.[11] envisaged solid-state processes occurring in ices on grains in very dense gas that is associated with high-mass star formation; these processes would be triggered by strong radiation fields from the newly-forming star. While the ices were cold (~10 K), a wide variety of radicals would be generated and retained in the ices. But as the temperature is expected to rise during the star formation process, mobilities of species within the ice mantles would also rise and simple radical association reactions would take place in which products could be stabilized by energy deposition into the ice lattice. The radicals generated in the relatively simple interstellar ices (as in Table 8.1) would obviously include OH, CH_3, CH_2, CH, NH_2, NH, CHO, CH_3O, CH_2OH, and many more. The potential products from the radical association reactions would therefore include, for example, molecules such as acetaldehyde, CH_3CHO, dimethyl ether, CH_3OCH_3, and ethanol, CH_3CH_2OH, ethylene glycol, $(CH_2OH)_2$, and very many more. In some cases, the reaction product would still be a radical, and so a second reaction stage would be necessary to create a molecule. For example, the association $CH + CH_2 \rightarrow CH_2CH$ (a radical) could be followed by the second stage $CH_2CH + CH_3 \rightarrow CH_3CHCH_2$ (propylene). The number of different product molecules from such a set of radical reactions could be potentially large, a hundred or more. Table 8.2 lists potential products from these radical associations. As the process of star formation continued, the grain temperature would also continue to rise, and for grain temperatures ~100 K, the ices would evaporate and the product molecules of this solid-state chemistry would be released to the gas phase on a relatively short timescale. It is evident that the possible products listed in Table 8.2 have many species in common with the detected species listed in the Appendix to Chapter 1 (when these molecules have at least one carbon and six atoms, in astrochemistry they are the so-called *Complex Organic Molecules*, or *COMs*).

Table 8.2 Possible products from radical association reactions in interstellar ices in very dense gas associated with the formation of a massive star. Named products are saturated molecules, and those shown in **bold** are species that have already been identified in interstellar clouds; those not bold have not yet been astronomically detected. Products that are not named are radicals that may undergo further reactions; see Table 8.3.

	OH	CH_3	CH_2	CH	NH_2	NH	CHO	CH_3O	CH_2OH
OH	**H_2O_2** Hydrogen peroxide	**CH_3OH** Methanol	CH_2OH	**H_2CO** Formaldehyde	NH_2OH Hydroxylamine	NHOH	**HCOOH** Formic acid	CH_3OOH Methyl hydro-peroxide	$CH_2(OH)_2$
CH_3		C_2H_6 Ethane	CH_3CH_2	CH_3CH	**CH_3NH_2** Methylamine	CH_3NH	**CH_3CHO** Acetaldehyde	**CH_3OCH_3** Dimethyl ether	**C_2H_5OH** Ethanol
CH_2			**C_2H_4** Ethylene	CH_2CH	CH_2NH_2	**CH_2NH** Methyl-enimine	CH_2CHO	CH_3OCH_2	CH_2CH_2OH
CH				**C_2H_2** Acetylene	$CHNH_2$	CHNH	CHCHO	CH_3OCH	$CHCH_2OH$
NH_2					N_2H_4 Hydrazine	N_2H_3	**NH_2CHO** Formamide	CH_3ONH_2 Methoxyamine	CH_2OHNH_2
NH						N_2H_2 Diazine	NHCHO	CH_3ONH	CH_2OHNH
CHO							$(CHO)_2$	**CH_3OCHO** Methyl formate	**CH_2OHCHO** Glycolaldehyde
CH_3O								$(CH_3O)_2$	**CH_3OCH_2OH** Methoxymethanol
CH_2OH									**$(CH_2OH)_2$** Ethyleneglycol

Table 8.3 Some examples of possible radical association routes to some detected interstellar species, using radicals produced in the association reactions listed in Table 8.2. Note that there may be several routes to the product species. All of these possible products are detected interstellar species. Named products are saturated molecules, and those shown in **bold** are species that have already been identified in interstellar clouds; those not bold have not yet been astronomically detected.

Radical 1 (from simple ices)	Radical 2 (from Table 8.2)	Possible product (all are detected species)
OH	CH_3CH_2	**CH_3CH_2OH (ethanol)**
OH	CH_2CH	**CH_2CHOH (vinyl alcohol)**
OH	CH_2CHO	**CH_2OHCHO (glycolaldehyde)**
OH	CH_3OCH_2	**CH_3OCH_2OH (methoxy methanol)**
OH	CH_2CH_2OH	**$(CH_2OH)_2$ (ethylene glycol)**
CH_3	CH_2OH	**CH_3CH_2OH (ethanol)**
CH_3	CH_2CH	**CH_3CHCH_2 (propylene)**
CH_3	CH_2CHO	**CH_3CH_2CHO (propanal)**
CH_2	CH_3CH	**CH_3CHCH_2 (propylene)**
CH_2	CHCHO	**CH_2CHCHO (propenal)**
NH	CH_3CH	**CH_3CHNH (ethanimine)**
CHO	CH_2OH	**CH_2OHCHO (glycol aldehyde)**
CHO	CH_3CH_2	**CH_3CH_2CHO (propanal)**
CH_3O	CH_2OH	**CH_3OCH_2OH (methoxy methanol)**

We noted in Table 1.3 that so-called "hot cores" are tiny clumps of very dense $[n(H_2) \sim 10^7 \text{ cm}^{-3}]$ warm $(T \sim 300 \text{ K})$ gas located close to newly-forming massive stars. These hot cores are locations where many of the detected organic molecules listed in the Appendix to Chapter 1 have been identified in the gas phase. Is it possible that these detected species are formed in solid-state chemistry and then released to the gas? If so, do these solid-state schemes operate in addition to gas phase schemes of the types discussed in Chapter 4 (if they exist), or do they replace them? Since a detailed theoretical investigation of solid-state schemes is very difficult because of the lack of reliable information about processes on ices of adsorption, desorption, photolysis, mobility, reaction probability, *etc.*, the best way to proceed is by designing suitable laboratory experiments involving realistic model ices under appropriate physical conditions.

8.3.2 Laboratory Studies of Chemical Complexity in Interstellar Ices

A number of important experimental investigations of various aspects of ice chemistry have been performed, several of which are mentioned here.

The chemistry induced by ultraviolet radiation in methanol ice as an analogue of ice in dense interstellar clouds confirms that a wide range of products can be formed.[12] In these experiments, the ices that have been used include those of pure methanol, or mixtures of methanol and either carbon monoxide or methane. The ices are grown under ultrahigh vacuum to thicknesses from 3 to 66 monolayers, and the temperatures are controlled accurately within the range of 20–200 K. The broadband irradiation is from a hydrogen microwave discharge lamp and has peak intensity at Lyman α (122 nm) and a wavelength range of 115–170 nm. Irradiation at these wavelengths leads to photodissociation of simple ice molecules in a single step; in Section 3.3, this process was defined as *photochemistry*. Detection of new species formed after irradiation is made using a Fourier transform infrared spectrometer in reflection-absorption mode (RAIRS). It was found in these experiments that some methanol was lost by photodesorption during irradiation, while some was converted to radicals which were available for chemistry (as suggested by Table 8.2). Irradiation was carried out at 20 K. A rich variety of photochemistry products was found and the rates of product formation from radical association were computed. Identified products formed in the ices included CO, CO_2, CH_4, HCO, H_2CO, CH_2OH, CH_3CHO, CH_3OCH_3, CH_3CH_2OH and $(CH_2OH)_2$. Further chemistry occurred during warming of the ices towards 70 K, indicating that diffusion of radicals through the ice is still occurring.

Energy sources other than ultraviolet are available in star forming regions, and the consequences of driving ice chemistry with these alternative sources have been explored experimentally. Young solar-type stars can be powerful sources of X-rays, and experiments have been carried out[13] in which methanol ice was irradiated with X-rays that were either monochromatic (300 eV or 550 eV) or broadband (250–1200 eV). Irradiation by X-rays is different in nature from that of ultraviolet irradiation: while ultraviolet is usually absorbed in a single event, the absorption of X-rays is much more energetic and occurs in a multi-stage process in which a succession of ionization and excitation events occurs; an initial high energy primary electron released in ionization generates multiple energetic secondary electrons that interact with the ice species. Ultimately, much of the energy appears as heat. This general process was defined in Section 3.3 as *radiation chemistry*. There are some differences in the products of radiation chemistry between the two monochromatic X-ray sources because the 300 eV photons tend preferentially to break C–H bonds while the 550 eV photons tend to break the C–O bonds. However, the list of products from ices energized by X-rays is – rather surprisingly – very similar to the list obtained from ultraviolet irradiation.

The effects on driving ice chemistry by energetic electrons[14] and by heavy cosmic ray particles[15] have also been considered. Both types of experiment produce ranges of products that are remarkably similar to – but slightly less extensive than – those produced by other forms of excitation, even though the basic mechanisms are likely to be very different.

Extensive laboratory studies covering many aspects of solid-state chemistry induced in ices have also been reported.[1,16] These investigations compare the chemistry resulting from both ultraviolet and X-ray irradiation, and they utilize several different kinds of ices. Interesting results arise when pure CO ice is irradiated; ultraviolet irradiation forms CO_2 and C_2O_3 and also causes CO photodesorption, whereas X-ray irradiation leads to the formation of a variety of products, including CO_2, C_3O_2, C_3, C_2O, either CO_3 or C_5 (these two are indistinguishable in the QMS detector), and C_4O. In spite of these differences, it is generally found that in most experiments, the products formed upon ultraviolet and X-ray irradiation are similar, although the relative abundances may differ. Results showing the molecules produced from either ultraviolet or X-ray irradiation of three different ice analogues are shown in Table 8.4. A wide variety is seen in these products, including many of those predicted in Table 8.2. Some COMs are also present: glycine (NH_2CH_2COOH) was detected in the residue of the ultraviolet-irradiated $H_2O/CO/NH_3$ ice, and other COMs including ethylene glycol (($CH_2OH)_2$), acetaldehyde (CH_3CHO), and formamide ($HCONH_2$) were observed.

The overlap between molecules formed in the solid-state chemistry of interstellar ice analogues and the molecules detected in interstellar space (in particular, in hot cores) is very striking. Whether or not there are gas phase routes to these COMs, it is clear that solid-state chemistry in interstellar ices driven by electromagnetic or particle irradiation is capable of producing many of these detected species. However, it should be noted that detections of COMs in interstellar and circumstellar regions are made using radio detections of *gaseous* species, while the experiments described above produce molecules embedded in the ices. Therefore, if – as seems likely – solid-state chemistry does indeed generate COMs in interstellar and circumstellar space, desorption of these molecules from the processed ices must occur. This desorption may be thermal or driven by the ultraviolet, X-ray, and particle irradiations that also drive the chemistry; X-rays may drive desorption more effectively than other means because of the much greater energy they deposit into the ice. Within the solar system, it may be possible to detect COMs desorbed from cometary nuclei. Glycine (NH_2CH_2COOH), the simplest stable amino acid, has been detected by the ESA-Rosetta mission to Comet 67P/Churyumov–Gerasimenko, and the presence of other molecules of prebiotic interest: ethylene glycol (($CH_2OH)_2$), acetaldehyde (CH_3CHO) and formamide ($CHONH_2$), has also been inferred.

The COM species formed in the experiments discussed here are relatively small and are formed with the input of external energy. It is unclear how these species can form molecules that are very much larger. However, recent work[17] indicates that a rich and complex chemistry of large chain molecules, peptides, may be achieved by the condensation of carbon atoms at 10 K without the assistance of an external energy source (Table 8.4).

Table 8.4 Molecular species identified in three types of ice analogue irradiated by either X-rays or ultraviolet. The identification is nearly always by infrared spectroscopy, and the wave number (in cm^{-1}) of the infrared band is shown; two species are identified by quadrupole mass spectrometry (QMS). The table shows that the primary products of the irradiation tend to be common in both types of irradiation, although the relative abundances may differ. Data are taken from ref. 1.

Ice	IR band (cm^1)	Molecular assignment	X-ray irradiation	UV-ray irradiation
CO	2348	CO_2	X	X
	2244	C_3O_2	X	X
	2039	C_3	X	
	1990	C_2O	X	
		CO_3/C_5 (QMS detection)	X	
		C_4O (QMS detection)	X	
H_2O/CO	1018	CH_3OH	X	X
	1096	($CH_2OH)_2$, CH_3OCH_3, C_2H_5OH, HCO, CH_3CHO	X	X
	1237	H_2CO, HCOOH	X	X
	1301	CH_4	X	
	1353	CH_3CHO, $HCOO^-$	X	X
	1382	$HCOO^-$	X	X
	1400	HCOOH, C_2H_5OH	X	X
	1499	H_2CO	X	X
	1583	$HCOO^-$	X	X
	1676	HCOOH	X	X
	1717	H_2CO, HCOOH, H_2CO_3, CH_3CHO	X	X
	1784	HOCO	X	X
	1852	HCO, HOCO	X	X
	2342	CO_2	X	X

(*continued*)

Table 8.4   (continued)

Ice	IR band (cm^{-1})	Molecular assignment	X-ray irradiation	UV-ray irradiation
H$_2$O/CO/NH$_3$ 2:1:1	2342	CO$_2$	X	X
	2261	HNCO, CH$_3$CN	X	X
	2242	HNCO, N$_2$O(?)	X	X
	2168	OCN$^-$	X	X
	2038	C$_3$, CH$_2$CNH, HNC	X	
	1848	HCO, HOCO, N$_2$O$_3$	X	
	1841	HCO, HOCO	X	
	1717	HCOOCH$_3$, H$_2$CO, HCOOH, H$_2$CO$_3$, NH$_2$CH$_2$COOH	X	X
	1690	HCONH$_2$	X	X
	1590	HCOO$^-$, CH$_3$NH$_2$, NH$_3$$^+CH_2COO^-$	X	X
	1495	NH$_2$, NH$_4$$^+$, NH$_3$$^+$, NH$_3$$^+CH_2COO^-$	X	X
	1478	NH$_4$$^+$, NO$_3$(?)	X	X
	1387	COO$^-$, HCONH$_2$	X	X
	1352	CH$_3$CHO, HCOO$^-$, NH$_3$$^+CH_2COO^-$	X	X
	1307	CH$_4$, N$_2$O$_3$, N$_2$O$_4$, HCONH$_2$(?)	X	
	1250	NH$_2$CH$_2$COOH	X	X
	1021	CH$_3$OH, O$_3$(?)	X	

References

1. G. M. Muñoz Caro, A. Ciaravella, A. Jiménez-Escobar, C. Cecchi-Pestellini, C. González-Díaz and Y.-J. Chen, *ACS Earth Space Chem.*, 2019, **3**, 2138.
2. F. C. Gillett and W. J. Forrest, *Astrophys. J.*, 1973, **179**, 483.
3. J. E. Chiar, Y. E. Pendleton and L. J. Allamandola, *et al.*, *Astrophys. J.*, 2011, **731**, 9.
4. J. M. C. Rawlings and D. A. Williams, *Mon. Not. R. Astron. Soc.*, 2021, **500**, 5117.
5. K. I. Öberg, *et al.*, *Astrophys. J.*, 2011, **740**, 109.
6. A. G. G. M. Tielens, *Molecular Astrophysics*, Cambridge University Press, 2021.
7. H. M. Cuppen, S. Ioppolo, C. Romanzin and H. Linnartz, *Phys. Chem. Chem. Phys.*, 2010, **12**, 12077.
8. T. Lamberts, H. M. Cuppen, S. Ioppolo and H. Linnartz, *Phys. Chem. Chem. Phys.*, 2013, **15**, 8287.
9. E. F. van Dishoeck, E. Herbst and D. A. Neufeld, *Chem. Rev.*, 2013, **113**, 9043.
10. N. Watanabe, O. Mouri, A. Nagaoka, T. Chigai, A. Kouchi and V. Pirronello, *Astrophys. J.*, 2007, **668**, 1001.
11. R. T. Garrod, S. Widicus Weaver and E. Herbst, *Astrophys. J.*, 2008, **682**, 2008.
12. K. I. Öberg, R. T. Garrod, E. F. van Dishoeck and H. Linnartz, *Astron. Astrophys.*, 2009, **504**, 891.
13. Y.-J. Chen, A. Ciaravella, G. M. Muñoz-Caro and C. Cecchi-Pestellini, *et al.*, *Astrophys. J.*, 2013, **778**, 162.
14. C. J. Bennett, S.-H. Chen and B. J. Sun, *et al.*, *Astrophys. J.*, 2007, **660**, 1588.
15. A. L. F. de Barros, A. Domaracka and D. P. P. Andrade, *et al.*, *Mon. Not. R. Astron. Soc.*, 2011, **418**, 1363.
16. A. Ciaravella, A. Jiménez-Escobar, G. M. Muñoz Caro, C. Cecchi-Pestellini, R. Candia, S. Giarrusso, M. Barbera and A. Collura, *Astrophys. J. Lett.*, 2012, **746**, L1.
17. S. A. Krasnokutski, K.-J. Chuang, C. Jager, N. Ueberschaar and T. Henning, *Nat. Astron.*, 2022, 381–386.

9 Interstellar Chemistry, Astrobiology, and the Origin of Life

"Here we report a demonstration that glycine, alanine and serine naturally form from ultraviolet photolysis of the analogues of icy interstellar grains." (M. P. Bernstein, J. P. Dworkin, S. A. Sandford, G. W. Cooper, L. J. Allamandola)[1]

"Most aspects of life can be understood rather well in terms of physics and chemistry, albeit an extraordinary form of chemistry that is highly ordered and organized, and of a sophistication that cannot be matched by any inanimate process." (Sir Paul Nurse)[2]

The last 13 billion years in the history of the Universe have been characterized by a persistent and inevitable increase in chemical complexity. As we have seen in Section 5.7.1, Big Bang nucleosynthesis occurred within the first few minutes in the existence of our Universe and gave rise to just three elements, hydrogen, helium and trace amounts of lithium. The first generation of stars was formed in this simple gas. Most of the 94 naturally occurring chemical elements are synthesized in stars. Once the usable nuclear fuel is exhausted, stars

Astrochemistry: Chemistry in Interstellar and Circumstellar Space
By David A. Williams and Cesare Cecchi-Pestellini
© David A. Williams and Cesare Cecchi-Pestellini 2023
Published by the Royal Society of Chemistry, www.rsc.org

die and expel their outer layers (see Section 6.2.2), thereby enriching the gas between them in new elements and their isotopes. In interstellar space, vast clouds of dust and gas originate new, increasingly metal-rich generations of stars, and the cycle repeats. Those elements that were produced and dispersed in the grand-scale cycle of matter into and out of stars provide the raw materials for planets in all their mineralogical diversity.

One of the second generation stars that formed in the Milky Way Galaxy 4.5 billion years ago is our Sun, and the chemical elements and their isotopes in our solar system are therefore the products of billions of years of Galactic chemical evolution. Chemical evolution on our planet (and perhaps in countless other planets and moons) led to the formation of biomolecules, and their assembly into organized molecular collections that eventually evolved into chemical systems able to self-reproduce in a truly genetic process involving both replication and translation (NB: for convenience, the Appendix to this chapter gives a glossary of some astrobiological terms used in this chapter). How then did these biomolecules originate and how did they come to be on Earth? These are uncomfortable questions, and the most honest answer should be: we don't really know. Life is perhaps the most sensational example of increasing chemical complexity, but the chemical evolution of the solar system and the planets within it represents an important precursor to life's origins and may provide a powerful key to understanding them. Wherever life begins, it must start from the kinds of molecules that are readily available. As we have seen in the preceding chapters, more than 250 different molecular species have been identified in interstellar space, and most of these are organic molecules. These molecules are involved in the processes that make stars, sticking to dust grains that combine to make meteoroids, planetesimals, asteroids, comets, and eventually planets. Alcohols, sugars and amino acids are detected in meteoritic and cometary material. Some of these organic species are observed in regions where solar-like stars and planetary systems are currently forming, and even in far-distant galaxies. Thus, it seems natural to wonder if these products of the routine cosmic chemistry, produced through the long history of biogenic elements, may have contributed to early Earth's organic pool, facilitating prebiotic molecular evolution. Moreover, life as we know it on Earth seems to be strictly related to homochirality in biopolymers. What is the origin of this peculiarity? Many different pathways have been proposed but the question is still unanswered. When considering the experimental evidence that the composition of various meteorites presents a fairly relevant excess of left-handed amino acids, we may also wonder if homochirality itself could be of extraterrestrial origin.

In this chapter, we discuss how a suite of quite simple molecular species with common organic functional groups, natural products of interstellar chemistry, can lead to more complicated biogenic compounds such as ribose, or a phosphate ester, or amino acids and nucleobases. Because these chemical processes are universal and some of the starting materials for life are likely to be widely distributed throughout the universe, we call this exercise *astrobiology*. There is something extremely interesting buried in this very fact. The connection to the origins of life is so elusive that we do not know what role exogenous prebiotic molecules may have played in implementing controlled chemical evolution. The origin of life may be described as a sequence of emerging events, each of which added new structure and chemical complexity, but it is highly unlikely that the precise evolutionary path in which these events took place can be specified. The challenge, therefore, is to establish a progressive hierarchy of chemical questions that deal with the particular mechanisms by which those underlying principles could have been expressed. Once such principles have been clearly outlined, the problem of abiogenesis would be transformed from understanding the essence into explaining the details. In this new perspective, the central role is played by chemistry. Biology is inseparable from its environment, but chemistry is indeed universal.

9.1   The Formation of Stars and Planets

According to our current knowledge, we can identify three fundamental episodes of chemical transition: firstly, stellar evolution and the emergence of elemental diversity; secondly, planetary evolution and the emergence of mineralogical diversity; and thirdly, prebiotic chemical evolution and the emergence of life. As we have seen in Section 5.7.1, the cycle of birth and death of stars constantly increases the abundance of heavy elements in the interstellar medium. These elements are dispersed into the interstellar gas or incorporated into micron-sized dust particles. Molecules like CO cool the interstellar gas much more efficiently than molecular hydrogen did for the primordial stars, triggering the formation of low-mass stars like our Sun.

The formation of stars and planets is among the topics of greatest interest in modern astronomy. Extensive surveys of star-forming regions in the Milky Way reveal a vast and intricate network of filamentary structures and dark bubbles, interspersed by bright hotspots where new stars appear. Filaments are nearly everywhere in the interstellar medium, giving rise to a universal web-like structure permeating the entire galaxy. These regions are typically in an approximate equilibrium between the forces acting to expand them and those trying to contract them, primarily gravity. In the densest strands, gravity takes over, and controls the fragmentation of filaments into *prestellar cores* and ultimately *protostars*. In this process, the gas is subjected to immense changes in its physical properties. The typical hydrogen number density within an interstellar cloud is a few hundred atoms per cubic centimeter, while its temperature is approximately 10 K. Turning these conditions into those of a star like the Sun requires an increase in number density by a factor of more than 10^{20}, and in central temperature by a factor of about a million.

If we ignore all processes inhibiting collapse, and imagine a cloud shrinking solely under gravity without resistance, the contraction takes place in the time required for all mass to fall to the central point, the so-called *free-fall time*. This is the minimum time necessary for the collapse to occur. It depends only on the gas density ($\sim 10^8 / \sqrt{n_{\mathrm{H}}}$ years). For a gas cloud with $n_{\mathrm{H}} = 10^4$ cm^{-3} or so, the free-fall time is of the order of one million years. Prestellar cores will later become the interior cores of stars. Over 50 000 years, the prestellar core contracts to a size of ~1000 AU. The process is violent and chaotic with the gas flowing in and being ejected outwards at speeds up to hundreds of kilometres per second, as the gravitational infall is locally opposed by thermal, turbulent, and magnetic pressures, by dynamical outflows, and – since the parent cloud is rotating – by angular momentum effects. As a consequence of all such competing processes, the contracting cloud forms a swirling disc, with the excess material stably ejected outward from the poles of the star to keep the system in balance. The system is now called a protostar, which means it is at its very first stage of becoming a real star. This continues for a further thousand years, until the star has grown enough in size and density for the central region to initiate nuclear reactions, which causes the star to shine.

The disc is now just a circular rotating plane of dispersed materials within which matter will slowly start to form clumps. These condensations grow bigger; they accrete more and more material and eventually form planets or moons. Eventually, the accretion ceases, the outflow weakens and the disc is largely eroded, although a remnant remains in orbit around the star. At this stage the disc has a size around 100 AU in diameter, and it is embedded into an immense, roughly spherical cloud (the so-called Öpik–Oort cloud) extending from about 20 000 to 100 000 AU, and consisting of icy bodies, relics of the planetary assembling. The construction of a planetary system is finally over, the disc is completely exhausted, and all the planets are formed. Over the next 10 billion years, the star will burn nuclear fuel in its centre (see Section 6.2.2), and emit energy as radiation. The outline given above is pretty much the same as the one for our own planetary system. Some phases of disc evolution are illustrated schematically in Figure 9.1.

High-mass stars evolve much faster than their low-mass counterparts, and the conditions in their vicinity are not conducive for planet formation, because the winds of massive stars carry away up to 10 billion times as much material as the solar wind at speeds of thousands of kilometres per second. So, even if planets do form around massive stars, they don't survive for long.

9.1.1 Chemical Heterogeneity in the Solar System

Approximately 4.5 billion years ago, our solar system was completely different from its present form. Instead of today's eight planets, there was a disc, composed mainly of hydrogen and helium, spinning around the newly created star. The early chemical history of the solar system was characterized by three phases. The first, the so-called *nebular phase*, occurred at the end of the collapse when the gas around the newly formed star was at its hottest, having been heated during the gravitational collapse.

In this first phase, the gas started to cool, although at the centre, the newly formed Sun began to shine, causing gradients in temperature and radiation. Since the vapour pressures of solids depend strongly on temperature, the raw materials available for planetary accretion varied strongly with the location in the disc. Close to the Sun, the materials were fully vaporized. As the temperatures decreased with increasing distance from the Sun, the gases began to interact chemically to produce new compounds in a process similar to dust formation in the expanding envelopes of cool stars (see Chapter 6). Chemical equilibrium determined the composition of solid materials in the early solar system, with baked and equilibrated solids residing in the relatively warm terrestrial planet region. Outwards, less and less heated and poorly equilibrated solids were present. The solar wind removed volatiles from the inner part of the disc, driving them towards the outer,

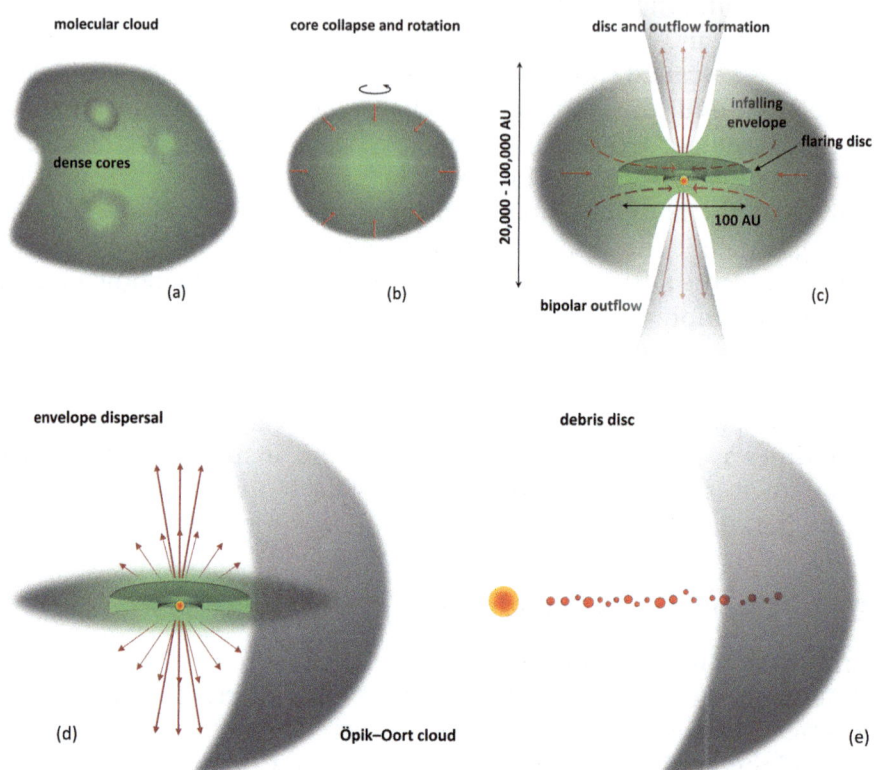

Figure 9.1 A sketch illustrating the formation of circumstellar discs and of planets within them. (a) Star formation begins in dense cores inside molecular clouds. (b) Sufficiently massive cores begin to collapse under their own weight. (c) Rotation and collapse of a core induce the formation of a circumstellar disc within the infalling envelope together with a fast outflow. (d) The outflow removes much of the infalling envelope, leaving a dense disc around the newly-forming star. (e) Dust settles on to the plane of the disc and accretes into clumps which may grow into planets and moons. Gaseous material is dissipated from the inner disc, while gas in the outer cooler disc may be trapped on outer planets.

much colder regions, until these species crossed an invisible boundary called the *snow* (or *frost*) *line*, where ices can exist without being melted or destroyed by the Sun's heat. Chemical equilibrium in the inner disc included successive condensation of refractory oxides and metals, metallic iron-nickel alloys, magnesium silicates, alumina-silicates, iron sulfide, and various types of oxygen-bearing (FeO, OH, and H_2O) silicates. In the process, chemical kinetics provided departures from equilibrium because of the inhibition at low temperatures of reactions such as the reduction of CO and N_2 to CH_4 and NH_3 (see Section 5.2).

In the outer regions, while species condensing at low temperatures may have approached equilibrium, high-temperature condensate grains inherited from the protostellar cloud would survive with little equilibration. These are "fossil", *i.e.*, presolar, solids originally formed in distant environments such as mass outflows from giant stars or nova and supernova shells (see Chapter 6), which are found to be quite unaltered in carbonaceous asteroids and comets. Thus, most of the materials were in equilibrium, whereas relic presolar solids were far from equilibrium with the bulk of nearby gas and grains. We find records of this history in meteorites.

The *second, accretionary, phase* of the chemical history of the solar system includes the physical and chemical processes associated with the assembly and growth of young planets. How discs of gas and rocks become planet factories is still under debate, but there is enough consensus that, in the beginning, dust grains (barely micrometres in size) stuck together to form fluffy solids. From that point on, accretion continued, producing in the terrestrial zone, the precursors of planets, called *planetesimals*, on a timescale of a few thousand years. The key steps in this assembly process are still controversial, but once planetesimals were large enough for self-gravitation to occur, planetesimals grew by collisions and mutual gravitational interactions, leading to the formation of a few tens of Moon-to-Mars-sized planetary embryos in roughly 0.1 to 1 million years. Full-sized planets then formed by large-scale collisions between these embryos within 10 to 100 million years. One of these collisions is thought to have led to the formation of Earth's Moon.

Planetesimals formed initially from a surprisingly small number of refractory condensed phases, such as corundum, spinel, graphite and silicon carbide. However, the chemical composition of planetesimals depended primarily on their initial distance from the Sun. Planets that formed close to the young Sun, such as the terrestrial planets, would be expected to contain less of the volatile icy and gaseous materials and thus be richer in the rocky materials. The most primitive materials are a class of meteorites called *chondrites* that appear to have not been significantly altered by melting or differentiation. One of the most striking features of these meteorites is the presence of *chondrules*, submillimetre to centimetre sized iron/magnesium/silicate spherules, remnants of molten droplets heated to near melting point and then cooled on timescales of hours. Even more primitive than chondrules are the *calcium–aluminium-rich inclusions (CAIs)*, found in chondrites. As indicated by their absolute radiometric ages obtained using the U-Pb chronometer, CAIs appear to be 4567.30 ± 0.16 million years old. They formed over a period of about 40 000 to 200 000 years. As CAIs are the oldest dated solids, this age is commonly used to define the age of the solar system. However, since the observed time span of stellar accretion (1–2 million years) is much longer than that taken by CAIs to form, it raised the question of how quickly the material that makes up the solar system was accreted. It is also unclear which phase in the solar system's formation is recorded by the formation of CAIs. Much of our understanding of the origin and evolution of the solar system depends on how well the timing and the processes that took place during this period can be constrained.

During and following their formation, terrestrial planets were subjected to catastrophic bombardments by the remaining rocky planetesimals. The impacts cratered the planetary surfaces, and – together with intense radioactivity and subsequent gravitational contraction – produced sufficient heat to melt and chemically differentiate the planets into their presently layered structure, namely *core, mantle*, and *crust*. The impacting material contributed, with inner heat sources, to the formation of the planetary atmospheres that emerged during this process and afterward by outgassing. This represents the third stage, the *bombardment phase*, in the history of materials in our planetary system.

In the region between Mars and Jupiter (the so-called asteroid belt), the temperature of the solar nebula was lower than in the region of the terrestrial planets, so carbon-rich and water-rich minerals could coalesce in the forming planetesimals. Further outwards from the asteroid belt, temperatures were even lower so other materials – such as ammonia and methane – were able to condense. Therefore, those more distant planetesimals had a composition consisting of rocky matter soaked by an icy mix of water, ammonia, and methane. Hydrogen and helium do not condense in the solar nebula, and these gases were rather abundant around objects in the outer solar system. As these outer planetesimals continued to grow larger, the strength of their gravity became more powerful. These bodies increased in size, growing to masses of a few times that of Earth. The surrounding material, primarily hydrogen and helium, was thus increasingly attracted, so the planetesimals accreted more and more material. This is the most likely route of formation of Jupiter and Saturn. Due to their great masses, these planets maintain the same relative proportions of hydrogen and helium to the heavier elements as are found in the Sun and in the interstellar medium. Uranus and Neptune are too massive to have been primarily outgassed, but they never underwent runaway gas accretion to the extent that Jupiter and Saturn did. Thus, carbon, nitrogen, oxygen, silicon and iron dominate their compositions. Comets are probably a probe of that distant era, being fossil relics of the primordial icy planetesimals that existed in the outermost regions of the solar nebula.

During their formation, giant planets underwent a violent, chaotic phase of orbital rearrangement under the combined influence of dynamical interaction with the gas disc and gravitational interactions with other planets, which caused the so-called *late heavy bombardment*, an intense spike in asteroid bombardment roughly 4 billion years ago, suffered by Earth and the entire inner solar system. This event shaped the terrestrial planet region. If life originated in environments on Earth's surface, it only could have happened after the bombardment stopped. Then the rapid and fierce evolution of the young solar system was replaced by the more regular secular evolution of the modern solar system.

9.1.2 Mineralogical Diversity

Meteorites are naturally occurring objects that originate on a planetary body and survive delivery to the surface of a different planetary body. They range in size from dust particles to bodies hundreds of meters across, with most of them being fragments of asteroids that broke apart long ago in the asteroid belt, located between Mars and Jupiter. The majority of meteorite finds are stony meteorites, consisting mostly of silicate minerals. Among them, chondrites, being some of the oldest materials in the solar system, show minimal structural changes compared with rocks from larger planets, which have been subjected to geological activity. The composition of chondrites reflects to some degree the composition of the solar nebula from which they formed. Carbonaceous meteorites contain a substantial amount of carbon (up to 5% by weight) and may exhibit evidence of thermal and aqueous alteration believed to have occurred on their parent bodies. They are rich in water-bearing minerals, organic compounds, magnetite (Fe_3O_4), and water-soluble salts. These objects are formed at low temperatures and contain the most complete inventory of volatile and moderately volatile elements. Chondrites with very low volatile and FeO contents have clearly a high-temperature origin. Their dominant silicate mineral is the FeO-poor silicate enstatite ($MgSiO_3$). Unaltered chondrites are characterized by a relatively poor mineralogical diversity, containing not more than 20 minerals. As planetary construction progressed, and chondrite parent bodies became larger, aqueous and thermal alteration led to new suites of minerals. Chondrites contain approximately 100 different mineral species.

Mineralogical diversity increased with the advent of asteroid melting and differentiation, processes that resulted in the large-scale separation of stony and metallic components. Those stony meteorites that have undergone melting, equilibration, and density-dependent differentiation are called achondrites, and they include meteorites from Mars and the Moon. They are igneous, meaning at some point they were formed when hot, molten rock crystallizes and solidifies. This occurs creating a concentric layered structure, in a process known as igneous differentiation. Differentiated meteorites include several classes of stony achondrites, as well as stony-iron and iron meteorites. As in chondrites, the mineralogical inventory is rather limited, containing about 300 different species (among 4000 minerals identified so far).

Rocky planets of our solar system suffered a similar fate. As the planetary material melted, heavy constituents sank to the bottom, whereas light ones floated to the top, resulting in the formation of distinct compositional layers. During the process, the minor and trace elements partitioned themselves between metals, sulfides, silicates and volatiles according to their chemical affinities. The accretion and rapid differentiation of Earth occurred about 4.55 billion years ago. Together with the Moon-forming impact 50 million years later, these processes produced a significant increase of near-surface mineralogical diversity, and led to the formation of a core, a crust, and eventually continents. The light elements were driven from the interior to form an ocean and atmosphere. Much of that diversification resulted from volcanic outgassing and fluid–rock interactions associated with the formation of the atmosphere and oceans. Interactions among the early atmosphere, the lithosphere and the hydrosphere produced hydrous silicates and oxides, including serpentinization – the hydrolysis and transformation of primary iron/magnesium minerals such as olivines $(Mg,Fe)_2SiO_4$ and pyroxenes $(Mg,Fe)SiO_3$ to produce H_2-rich fluids, and a variety of secondary minerals. These processes combined with the first significant production of clay minerals, and perhaps zeolites, namely hydrated aluminosilicate minerals that contain alkali and alkaline-earth metals.

Throughout Earth's history, mineralogical evolution has been driven by cyclical selective processes: heating and cooling, melting and crystallization, burial and uplift, weathering and sedimentation, dissolution and precipitation. All new minerals arose from natural selective processing of older minerals, although the last billion years were characterized by a boost in mineralogical diversity as a consequence of atmospheric oxygenation, the imprinting of a new chemical event, the emergence of life.

9.1.3 Organic Chemistry in Meteorites

About 85% of meteorites that presently fall on Earth are chondrites. Carbonaceous chondrites representing 5% of the whole group of chondrites are considered the most primitive, and contain carbon in several forms as silicon carbide, graphite, nanometre-sized diamonds originating as condensation products of stellar outflows as well as from catastrophic supernovae explosions that occurred long before the birth of the Sun (see Section 6.3), and younger carbonate minerals. However, all of these forms of carbon are less abundant than carbon in organic compounds.

A large body of detailed chemical analyses found that the majority, about 80%, of this organic matter is in the form of an insoluble macromolecular material, something like terrestrial kerogen (the primary organic component of oil shale) consisting mainly of paraffin hydrocarbons. This highly cross-linked aromatic network or macromolecular material has the average elemental abundance $C_{100}H_{46}N_{10}O_{15}S_{4.5}$. Organic chemists give the name *tars* to these hydrogen-poor substances. The rest of the carbon is a soluble complex mixture of compounds, in which the most notable (but not the most abundant) components are amino acids, nucleobases, and sugars, all of which are fundamental to life on Earth. There have also been identified other free molecules such as hydrocarbons, carboxylic acids, hydroxy acids, sulfonic acids, phosphonic acids, poly-hydroxyl compounds and many other chemical species that are relevant to the origins of life. However, the majority of the more than 100 different amino acids identified in carbonaceous meteorites are non-existent, or rare, in terrestrial proteins. A list of the classes of organic compounds found in the iconic Murchison meteorite is reported in Table 9.1.

9.1.4 Planetary Atmospheres and Moons

There are two broad types of planetary atmospheres, called primary and secondary. The gas giant planets have primary atmospheres consisting of gas captured from the protoplanetary disc, and composed

Table 9.1 Organic compounds in the Murchison meteorite.

Class	Example	3D structure
Carboxylic acids	Acetic acid	
Amino acids	Alanine	
Hydroxy acids	Lactic acid	
Keto acids	Pyruvic acids	
Dicarboxylic acids	Succinic acid	
Sugar alcohols and acids	Glyceric acid	
Aldehydes and ketones	Acetaldehyde	
Amines and amides	Ethylamine	
Pyridine carboxylic acid	Picolinic acid	
Purines and pyrimidines	Adenine	
Alyphatic hydrocarbons	Propane	
Aromatic hydrocarbons	Naphthalene	
Polar hydrocarbons	Isoquinoline	

mainly of molecular hydrogen, helium, and lesser amounts of ammonia and methane. The terrestrial planets, Venus, Earth, and Mars in our solar system, either lost or never had primary atmospheres. They possess secondary atmospheres derived from volatiles outgassed from their rocky interiors. The atmospheres of Venus, Earth, and Mars are oxidizing, comprise a very small fraction (lower than 0.01%) of the planets' total masses, and exhibit sharp boundaries with their surfaces. By contrast, the reducing atmospheres of the giant planets are consistent parts of the total planetary masses, and appear to extend deep into the planets, reaching extremely high pressures without observable solid surfaces.

Jupiter, the largest planet, is predominantly composed by H and He, with an outer gaseous layer surrounding a dense core of rocks and ice. Like Jupiter, Saturn is mostly made up of hydrogen and helium that surround a dense core. Its atmosphere is rather similar to Jupiter's. At the temperatures and pressures deep inside these giant planets, hydrogen becomes so heated and compressed that it enters several strange states, including liquid metallic hydrogen. Hydrogen is a non-metal, but under the extreme conditions in the planet's interior, atoms actually lose their electrons, creating a mix of protons and electrons. Because the electrons are unbound, they can move easily between the nuclei, a property associated with metals. Such an electrically conducting fluid is responsible for dynamo currents powering Jupiter's and Saturn's magnetic fields. Uranus and Neptune present outer atmospheres resembling those of Jupiter and Saturn and rocky cores, but they are too dense for hydrogen to be a major component of their makeup. Their inner regions are made up of about 65% water and other so-called "ices" (it is more a dense fluid of "icy" materials) such as methane and ammonia. The hydrogen inside these planets is never subjected to the temperatures and pressures required to reach a metallic state, and their magnetic fields are probably generated by dynamo currents in ionic oceans of aqueous ammonia occurring in the depths.

The primary difference among these planets is that the atmospheres get progressively cooler with increasing distance from the Sun. The chemistry in the atmospheres of these giants is determined by their overall elemental composition, and the stability of chemical compounds at the prevailing temperatures, the gravity-controlled total pressures, and in the upper atmosphere by the interaction with ultraviolet solar radiation and other energy sources (*e.g.*, cosmic rays). Deep into the atmosphere the chemical composition is at equilibrium, while in the upper regions, photochemistry powers drastic departures from equilibrium chemistry. Many of the gases observed in their atmospheres are hydrides, *e.g.*, CH_4, NH_3, H_2O, H_2S, PH_3, GeH_4, and AsH_3. These thermodynamically stable forms may be photo-destroyed in the stratosphere by UV photons and converted into, *e.g.*, ethane (C_2H_6), acetylene (C_2H_2), and ethylene (C_2H_4) from methane, and hydrazine (N_2H_4) from ammonia. Each planet has distinct cloud layers, with altitudes dictated by the atmospheric levels at which various gases can condense into liquid droplets or solid flakes. In Jupiter and Saturn, ammonia is removed by formation of NH_3 ice clouds. In Uranus, water vapor and H_2S condense to form clouds of liquid water and ammonium hydrosulfide (NH_4SH). The highest layers of Neptune contain cirrus clouds made up of frozen methane. Neptune also contains a haze at very high altitudes made up of hydrocarbons.

The solar-like composition of primary atmospheres was rapidly lost by planets in the inner solar system. This occurred through a combination of surface temperature, the mass of the atoms and their escape velocity from the planet. Since the gaseous light elements hydrogen and helium were lost, the remaining elements occurred in rocky (iron, olivine, and pyroxene) and icy materials (H_2O, CO_2, CH_4, NH_3, and SO_2). These latter materials, common in the outer regions of the planetary system, were delivered by comets, and mixed together with the rocky component in the early crust and mantle. The subsequent history depends on how quickly a planet loses its heat. If the planet cools quickly, there is little tectonic activity and the icy materials are trapped in the mantle. The larger inner planets, such as Venus, Earth, and Mars, are big enough to have remained hot inside, and still have active tectonism and volcanism. The more volatile materials are turned to gases, and returned to the planet surface to produce a secondary atmosphere. These outgassed atmospheres are very thin compared to primary atmospheres. The composition of the outgassing was similar for Venus, Earth and Mars, and consisted of H_2O (60%), CO_2 (25%), SO_2 (10%), N_2 (5%), and traces of noble gases (Ne, Ar, Kr). The current atmospheric and surface conditions of these planets arose as the result of a chain of astrophysical, geophysical, and chemical processes, including additional delivery of volatile-rich chondritic materials. Thus, these planets evolved in very different ways. The end results were the formation of a habitable environment on Earth, and highly inhospitable conditions on Venus and Mars.

Primary agents governing the evolution of an atmosphere were likely to be the surface temperature and chemistry of the planet. Water is the key catalyst for the evolution of a secondary atmosphere. Earth was nestled in our solar system at just the right distance from the Sun for liquid water to exist. The CO_2 released by outgassing was dissolved in this liquid water to produce carbonate rocks. Venus was too hot, and therefore the process did not occur and CO_2 remained as the primary component of its atmosphere. Mars experienced a period of liquid water very soon after formation, but due to the insufficient temperature for this water to remain as a liquid, it froze out leaving CO_2 as the primary component in the atmosphere. The atmospheres of Venus and Mars are today primarily made of CO_2 with traces (a few percent) of N_2. Thermochemistry is the dominant process controlling atmospheric composition on Venus due to the high surface temperature and pressure, whereas photochemistry dominates in the cold, thin atmosphere of Mars. If CO_2 disappears, lesser abundant elements become important. As a consequence, Earth is significantly different from either Venus or Mars with an atmosphere composed primarily of N_2 with later additions of O_2 from lifeforms. The composition of Earth's atmosphere is controlled by biological processes, with photochemistry also playing a significant role, particularly in the generation of the ozone layer. The atmospheres of all three planets possess clouds, and circulation in response to thermal forcing by the Sun. On Earth, clouds generally form within the troposphere, obscuring just a fraction of the sky, but on Venus the clouds never part, for the planet is enveloped entirely in a 20 km-thick veil of carbon dioxide and sulfuric dioxide haze, together with thick clouds composed mainly of sulfuric acid droplets. Clouds on Mars are relatively rare, thanks to the planet's thin, dry atmosphere. They consist of water ice condensed on reddish dust particles suspended in the atmosphere.

There are over 150 moons orbiting the solar system's planets. Of these moons, some have known atmospheres. The Moon (orbiting Earth), Ganymede, Europa, and Callisto (Jupiter), Rhea, Dione, and Enceladus (Saturn), and Titania (Uranus) have extremely thin atmospheres. Most of these atmospheres contain some mixture of oxygen, methane, nitrogen, and carbon monoxide and dioxide. The Moon shows the presence of some unusual gases, including sodium and potassium, which are not found in the atmospheres of Earth, Mars or Venus. Primary components are neon, helium, and hydrogen, in roughly equal amounts. There are also minor concentrations of methane, carbon dioxide, ammonia, and even water. Ganymede is the ninth-largest object (including the Sun) of the solar system, and it is the largest without a substantial atmosphere. Observations have found evidence of a thin oxygen atmosphere, thought to originate from Ganymede's icy surface (daytime surface temperatures range from 90 to 160 Kelvin). Europa presents an ice shell 15 to 25 kilometres thick, floating on an ocean 60 to 150 kilometres deep. Although Europa is only one-fourth of Earth in size, its ocean may contain as much water as in all of Earth's oceans. Europa has only a tenuous atmosphere of oxygen, but appears to be actively venting water into space. Saturn's moon Enceladus joins Titan (the largest moon of Saturn), Jupiter's moon Europa, and Earth in being a solar system body with liquid water on or below its surface. An ocean of water lies beneath the icy surface of Enceladus, at a depth of about 30 kilometres. Near that moon's south pole, jets of water escape into space, possibly surfacing through cracks in the ice shell. Both Triton (Neptune) and Io (Jupiter) have more substantial atmospheres, although still 100 000 times thinner than Earth's atmosphere. Both of these moons experience weather, seasons, and even clouds. The atmospheres of Triton and Io are primarily composed of nitrogen and SO_2, respectively. Lastly, Saturn's moon Titan is unique in being the only moon with a thick atmosphere, about 50% denser than Earth's. It primarily contains nitrogen and methane, and has weather, cloud formations, seasons, and perhaps lightning. Icy moons orbiting gas giants provide an intriguing alternative to Earth-like planets for the search of life in the universe.

9.1.5   Exoplanets

Exoplanets are planets beyond our own solar system. Planets are billions of times fainter than their parent stars, so they were thought to be unobservable. In the last two decades, however, astronomers successfully developed indirect detection methods, based on the effect that planets have on the luminosity of their parent stars. The first exoplanet was discovered in 1992 and, since then, more than 8000 exoplanet candidates have been observed, and almost 5000 are considered to be confirmed.[3] These worlds come in a huge variety of sizes and orbits. Some are gigantic planets hugging close to their parent stars, others are icy, some rocky.

The methods used to detect exoplanets are simple in principle, although extreme precision is required in order to detect minimal effects on the much larger star. Accuracy may also be reduced by distortion due to Earth's atmosphere, so space missions are becoming the norm in these observations. The methods we refer to are indirect, opposite to the straightforward approach to isolate either the light being reflected by the exoplanet or the thermal infrared radiation emanating from the planetary surface itself. These direct techniques are strongly limited by the extreme brightness of the parent star compared to the emission from the planet we are trying to detect. The indirect methods are based primarily on motion and obscuration of the parent star caused by the orbiting planet. As both the star and the planet orbit their common centre of mass, the slight motion back and forth of the star causes the stellar spectral lines to be displaced due to the Doppler effect. In addition, if Earth lies in the plane of the planet's orbit, then when the planet passes directly in front of its star, the starlight is slightly dimmed. This tiny reduction can be recorded, after observing the stellar emission over a period of time. In the first case, the oscillations of the star give information on the planet-to-star mass ratio, while the weakening of the stellar emission depends on the size of the transiting object.

Astronomers classify the various types of exoplanets as gas giants, hot Jupiters, super-Earths, water worlds, and exo-Earths, these names making up just a list of oddities. Hot Jupiters are gas giant planets with orbital periods less than 10 days. The short period means that hot Jupiters are very close to their host stars, usually less than 0.1 AU, one tenth of the distance between Earth and the Sun. Since they are incredibly close to the host star (at least with respect to planets in the solar system), they are extremely hot. These planets are places where iron gets vaporized, condenses on the night side and then falls from the sky like rain. A super-Earth is a rocky planet bigger than Earth, while a water world is a type of planet predominantly covered in water.

Exoplanets can be divided into three groups, terrestrials, gas giants, and mid-sized gas dwarfs. There are two clear dividing lines: one between 1.5 and 2 times the radius of Earth, and the other at 4 times larger than Earth, which appear to mark changes in composition. Smaller planets are likely to be rocky, while larger ones are probably gas giants. In the middle, the rocky cores of gas dwarfs formed early enough to accrete some gas, although they were unable to grow as large as gas giants like Jupiter. Combining data from the two different observational techniques, it is a simple matter to compute the average density of an exoplanet dividing the mass by the volume. Density, in turn, can suggest whether a planet is rocky, gaseous, or in between, and whether or not it might have a significant atmosphere. For example, for medium to large-sized planets which are not very dense, it is possible to determine that they are likely to have an atmosphere, or perhaps that their composition is mostly ice and gas. On the other hand, dense planets are more likely to be rocky, and some may be composed of more exotic materials such as graphite or even diamond. Some of these planets might also resemble the watery moons around Jupiter and Saturn. If a planet's density is lower than Earth's, that is an indication that the planet could host more water and not as much rock and iron, and if the planet's temperature allows liquid water to exist, we may have an ocean or water world. For lower densities and temperatures, the planet could harbour an icy ocean.

As a planet transits its central star, some light will go through its atmosphere. That light can be analyzed to determine what different atmospheric elements influenced that particular dispersion. The compositions of extrasolar planetary atmospheres may help to understand the type of planet, the processes operating on the planet, and whether or not the planet could support life. Exoplanet atmospheres are related to their interiors, but how close this relationship is remains an open (and difficult) question. The expected atmospheres may be categorized with respect to the presence or absence of volatiles; in our own planetary system, this represents a well-defined dividing line:

1. prevalence of H and He; these elements indicate capture of gas from the protoplanetary nebula (*e.g.*, giants and ice giants in the solar system);

2. outgassed atmospheres with hydrogen that has not been captured from the nebular disc; this requires a planetary mass substantially larger than Earth's and temperatures low enough to retain hydrogen against atmospheric escape; the atmospheric composition should be different from that of solar system planets with CO_2 or N_2 dominated atmospheres; helium should not be present, since it cannot be trapped in rocks or accreted during formation;

3. outgassed atmospheres that have lost both hydrogen and helium, and are CO_2 dominated; if the planet is water rich (like Earth), CO_2 is removed leaving N_2 as the dominant atmospheric gas;

4. hot super-Earth atmospheres that have lost not only light elements but also other volatiles such as C, N, O, and S; the atmosphere would then be composed of silicates enriched in more refractory elements such as Ca, Al, and Ti; at very high temperatures, the surface would be totally molten having thus oceans of lava (lava planet), and the silicates in the atmosphere would condense into clouds;

5. planets without an atmosphere or with a minimal one (*e.g.*, Mercury and the Moon).

To date, most atmospheric detections have been made in hot Jupiters or hot Neptunes that orbit very close to their star, and thus have heated and extended atmospheres. H_2 is the dominant constituent of most exoplanets. However, alkali species such as sodium, potassium and lithium were the first constituents detected in the atmospheres of hot Jupiters. CO is also abundant in these environments and water is roughly as abundant as CO, while CH_4 is not frequently observed. Other observed molecules are HCN, NH_3, C_2H_2, CO_2, and OH, this latter species being a product of the destruction of water vapour due to the extremely high temperature. Finally, the presence of metal oxides (*e.g.*, TiO and VO) and hydrides (*e.g.*, FeH, CrH, TiH, and ScH) is unclear.

Molecular observations of exoplanet atmospheres are challenging given the stringent requirements on the sensitivity. Moreover, the presence of clouds flattens spectral features in the transmission spectrum, because of the inability of stellar photons to reach depths in the atmosphere below the cloud top caused by the scattering and absorption of radiation by the particles forming the cloud. Images of the worlds in our solar system that have atmospheres show clouds or haze as either local or global features. Transmission spectroscopy has revealed that clouds and hazes are common also in the atmospheres of exoplanets. Chemical analyses suggest that these clouds may be composed of exotic materials, such as salts, sulfides, rocks, metals, and hydrocarbon "soots", although in practice many of these terms are imperfect, as aerosols often clump together to form complex mixtures.

Being a universal solvent, water is thought to play many roles in supporting life. As a fundamental requirement of living systems, water holds a special place in the conceptual framework of astrobiology. All of life's processes are carried out in the presence of liquid water, and on this basis, it may be regarded as a key indicator for potential habitability. To qualify as potentially life-friendly, a planet must be relatively small, and therefore rocky, and orbit in the "habitable zone" of its star. This zone is loosely defined as a region around a star where the conditions could potentially be suitable to allow for the presence of liquid water on its surface. However, water molecules not only serve as a solvent and reactant but can also promote hydrolysis, which counteracts the formation of essential organic molecules. While inside cells hydrolysis is a controlled reaction, how could it have been regulated under prebiotic conditions? Formamide ($HCONH_2$) may have been early life's alternative to water. Formamide is made by the hydrolysis of hydrogen cyanide, which is in turn formed from hydrogen, carbon, and ammonia. This species is not only found on Earth, but also in star-forming regions, within interstellar clouds, and in comets (see the Appendix to Chapter 1).

9.2 Astrobiology

Astrobiology aims to find the origin and distribution of life in the universe. This implies an understanding of life and the nature of the environments that support it. However, arriving at a working definition for life has historically proved to be difficult. Equally challenging is trying to understand how life may be detected in the Universe beyond Earth, primarily by identifying habitable planets. This idea generates the notion of an astronomical *biosignature, i.e.: what constitutes evidence for life?* Life perturbs disequilibria that arise due to kinetic barriers, producing chemical signatures that would not be expected from abiotic processes alone. As a consequence, a biosignature specifically requires a biological agent. This is a broad definition and one that may unfortunately be rather misleading because our concepts of life and biosignatures are inextricably linked. However, since nothing better can easily be found, we assume a biomarker to be a measurable indicator of some biological state or condition, such as cellular components or isotopic fractionation. A biomarker is also, by definition, an indication of habitability.

Biosignatures must reflect fundamental and universal characteristics of life, and they should not be restricted solely to those attributes that represent local solutions to the challenges of survival. For instance, certain specific components of our biosphere, DNA and proteins, might not be necessarily mimicked by other examples of life elsewhere in the cosmos. On the other hand, *e.g.*, amino acids appear to be natural products of cosmic chemistry. Thus, basic evolutionary principles might indeed be universal, as key molecular, or more organized, structural events that link a specific environmental exposure to a complex outcome. Again, this definition, even if restricted enough, encompasses a very large domain, and it opens out to the more general question of how biosignatures can be distinguished from the manifestations of the abiotic world. We need to define life in general measurable terms, considering that biosignatures are present over various spatial and temporal scales, which means, in such research, to consider molecular structure, chemistry, replication, energy budget, and environmental conditions. Thus, biosignatures should also include atmospheric gases, minerals and all those recordings of the impact on a planetary scale of the all-pervasive conflict between entropy and enthalpy (in a sense that would be very familiar to J. W. Gibbs).

Since we know that life is strongly affected by environmental conditions, the search for exoplanets may be useful in identifying unexplored and extreme physical and chemical conditions which differ substantially from those of terrestrial chemistry. A chemistry exhibiting a marked departure from thermodynamic equilibrium is necessary, but hardly sufficient. For instance, byproducts of metabolism are often thought as the most promising biomarkers, because they are characterized by extreme thermodynamic disequilibrium, as in the classic case of the simultaneous presence of O_2 (or O_3) and a reduced gas such as CH_4. However, there is a long list of sources that could create either true, *e.g.*, impacts, or apparent, *e.g.*, composite emission spectra, disequilibrium in a planet's atmosphere, without implying the existence of extraterrestrial life. In principle, we may try to identify a bounded mechanism in which an external flow of molecules and energy produces components which, in turn, continue to maintain the organized bounded structure that gives rise to these components. An abiotic process invariably results in products that are foreign to the process itself. Life is not just a highly ordered system, but it can also reproduce, and through this process, it can maintain itself. Moreover, with the reproduction, it gains new properties and adapts to new environments. This is an important point: autocatalytic cycles with molecules acting as templates do not meet the minimum requirements for life because they can't evolve. Although this may sound terribly abstract, a familiar realization of processes that create themselves, sustain themselves, and produce themselves is the cell, the basic structural and functional unit of all known living terrestrial organisms. Moreover, life adapts to changing environments, and also changes the environments. In this way, instead of extrapolating local structures to alien environments, terrestrial-based biosignatures should be enclosed into a universal class of phenomena.

A possible approach could be to exploit terrestrial self-maintaining chemistry in biological systems as a guide from which to construct recipes for alien systems in terms of their organization rather than the specific components. A remarkable example is the discovery in a warm interstellar cloud towards the galactic centre of the species iso-propyl cyanide (see the Appendix to Chapter 1), a molecule in which the carbon structure branches off in a separate strand, a fundamental characteristic of amino acids. The structure of iso-propyl cyanide may be written: $(CH_3)_2CHCN$. Given the high abundance of this species, it appears that branched molecules may in fact be the rule, rather than the exception, in space.

9.2.1 Astrobiology and Analytical Chemistry

We can hope to constrain such operational definitions of a biomarker, and build a protocol that works, approaching the problem in the laboratory, in the field on Earth, and in solar planetary studies, from ground- or space-based telescopes, flyby or orbit, to *in situ* investigations. In an origin-of-life context, it is important that these environments be well-characterized. To this aim, we need to turn to analytical chemistry, the measuring science that studies the chemical constituents, contents, distribution, and interactions of matter. Observing the atomic and molecular constitution of distant galaxies or analyzing samples taken from asteroids and planets requires rather specialized analytical methodologies and instrumentation.

9.2.1.1   Analytical Techniques

Gas chromatography is a widely used technique in analytical chemistry. The method consists of the heating and vaporization of the compounds contained in the sample. A chemically inert gas serves to carry the molecules of the analyte (the sample being analyzed) through a separation tube known as the "column", which allows the sequential elution (separating different materials) of separated components. The combination of gas chromatography and *mass spectrometry* is an invaluable tool in the identification of molecular species. Mass spectrometers are instruments used to measure the mass of electrically charged particles, these particles being affected by a magnetic field that provides a force perpendicular to their motions. Thus, the first step in the measurement is knocking one or more electrons off a molecule to leave a positive ion. The ion is then deflected by the magnetic field according to its mass. The more the ion is charged, the more it is deflected. What a mass spectrometer actually measures is the mass-to-charge ratio of the ions. Mass spectrometer detectors have sensitivity ranges as low as 1–10 nanograms or 1–10 picograms for some species. *Chromatography* coupled to mass spectroscopy can be used to determine atmospheric compositions, or the components of samples collected on planetary surfaces. This technique can detect volatile organic compounds, hydrocarbons, amino acids, porphyrins, organic acids, alcohols, aldehydes, and ketones, all of which can be strong indicators of biological activity.

As we have seen in Chapter 2, *spectroscopy* measures how matter interacts with electromagnetic radiation. Every molecular species has a unique way of interacting with radiation. These interactions can be used to detect many organic and inorganic compounds in or on planetary bodies using *in situ* or remote sensing instruments. Spectroscopy can be used in absorption, emission, and scattering modes. In the ultraviolet band, we may obtain the chemical composition, densities, and temperatures of the interstellar medium and stars, identify the surface composition and structure of atmospheres, determine volatile gas loss rates, and search for ice surface layers. Near-infrared spectrometers map morphological features, while simultaneously determining their composition and mineralogy, and providing data to investigate the evolution of surface geology. In very thin atmospheres, such as that of Mars, tunable laser spectrometers can be used to measure concentrations of gases such as methane, carbon dioxide, and water vapour. X-ray radiation can interact with the electrons of the inner shell of an atom, providing information on the chemical and elemental properties of a sample and allowing for the analysis of materials inaccessible to other techniques. In Raman spectroscopy, scattered light is used to measure the vibrational energy modes of a molecular sample. It can provide both chemical and structural information, as well as the identification of substances through their characteristic Raman fingerprints.

Wet chemical analysis is another form of analytical chemistry that refers to chemistry performed on samples in the liquid phase. For quantitative purposes, this method may use the procedure of decomposing a sample (with reagents such as acids) to dissolve in a solvent and identifies and quantifies the targeted elements using various measurement methods. The NASA *Phoenix* mission, the sixth successful landing on Mars, carried out the first wet chemistry experiments on another planet. An example of a wet chemistry instrumental technique is the so-called ion-selective electrode, a transducer, or sensor that converts the activity of a specific ion dissolved in a solution into an electrical potential.

Finally, *nuclear magnetic resonance* is an analytical chemistry technique used for determining the content and purity of a sample as well as its molecular structure. Many nuclei have spin and all nuclei are electrically charged. If an external magnetic field is applied, an energy transfer is possible from the base energy to a higher energy level. This takes place at radio frequencies, and when the spin returns to its base level, energy is emitted at the same frequency. A device using this method measures this energy and yields a spectrum for the nucleus concerned.

9.2.1.2 Mars

Mars has a unique place in planetary exploration, and presents an excellent opportunity to investigate the major question of habitability and life in the solar system. Once similar to our planet, Mars is now a cold desert, wrapped in a thin CO_2 atmosphere. At some point during the evolution of Mars, a dramatic event transformed this formerly Earth-like world into the dusty, dry husk we see today. Thus, by exploring Mars we may learn about variations in climate that can fundamentally alter planets. Moreover, we may search for signs that might reveal whether life was abundant in the planet's past, and if there are still traces on Mars today. Although unlikely, life may yet exist today on Mars in some protected subsurface environments, shielded from the harmful effects of ionizing radiation, reactive chemical oxidants and desiccation.

The spacecraft exploration of Mars began in the 1960s with a strategy of flybys, followed by orbiters, landers, and rovers with some kilometres of mobility. This systematic investigation has produced a detailed knowledge of the planet's character. Through remote spectroscopy, the Martian surface has been mapped in search of aqueous or hydrothermal activity, which has implications in determining past habitability and the potential for past or present life on the planet. In addition to phyllosilicates (a hydrous mineral), and olivine, carbonate-bearing rocks have been detected. All of these minerals were likely produced by aqueous alteration indicating the existence of liquid water in the past. These detections imply that the planet may once have had a vast ocean covering its northern hemisphere. Surface missions have confirmed that water has played a major role in the history of the planet for long periods. There is also evidence of past rainstorms, and ancient lakes and rivers that carved troughs into the terrain. Mars also likely hosted a thick atmosphere capable of maintaining liquid water at Martian temperatures and pressures.

Of critical importance is the geological record of early Mars that is largely lost on our own planet. This crucial early period is when life began (at least on Earth), and therefore there is the potential for evidence of prebiotic and biotic processes and how they relate to the evolution of the planet as a system. *In situ* organic chemical analyses have been conducted over the past four decades using multiple generations of technology, aimed at ascertaining the chemical and isotopic composition of the Martian atmosphere, and analysing volatiles extracted from solid samples, to assess quantitatively the habitability of Mars. The results of direct tests have produced a suite of organics including thiophenic [thiophene is the five-membered ring $(CH)_4S$], aromatic, and aliphatic compounds. In some cases, a wet chemistry technique has been employed to ease the detection of organics with the gas chromatograph and mass spectrometer: organic molecules are "derivatized" before they're heated, through reactions with other chemicals, in order to make them more volatile. The observed atmospheric methane abundances are generally low, while evidence of spikes in methane levels remains ambiguous, because of the spectral overlap of important methane features with ozone spectral features. Since methane is relatively short-lived in an oxidizing atmosphere, its presence (and variations) is considered a biomarker, and one possible piece of chemical evidence of ongoing microbial metabolism.

9.2.1.3 Water Worlds

Other worlds are beginning to be the subject of dedicated space missions. Going forward in our search for life, some of the most enticing targets in the solar system for these investigations are those worlds that contain liquids, heat, and nutrients. These include the Jovian companion Europa and the saturnian satellites Enceladus and Titan, where demonstrably habitable environments are most likely to be found. They are thus becoming increasingly attractive targets for astrobiology. These bodies with their subsurface liquid water oceans under their icy surface are likely to harbour hydrothermal activities, similar to the sub-surface vents in terrestrial oceans. In Enceladus, warm salt water, gases and minerals erupt through fractures in the ice shell and are ejected at more than 1000 km per hour into space (these geyser-like events are called plumes; see Figure 9.2), where spacecraft instruments can analyze them.

Using state-of-the art analytical detection techniques with very high sensitivity, some evidence has been found that suggests that the global ocean on Enceladus hosts active hydrothermal vents. These

Figure 9.2   Plumes on Enceladus. Water jets from Enceladus' ocean escape into space through cracks in the ocean's ice shell. The image was captured by NASA's Cassini spacecraft. Credit: NASA/JPL-Caltech/Space Science Institute.

environments were previously known to exist only on Earth. They present extreme conditions, but also eject heat and nutrients, and may therefore be considered as potential sites for the emergence of life (see Section 9.3.1). Water from the plumes of icy moons is a potential sampling target that may preserve the aqueous chemistry of the interior oceans, as well as organics and other prebiotic chemical precursors. The search for biogenic elements and organic compounds, and thus potential biosignatures, has stimulated various missions to these icy moons. Anticipated missions over the next few decades by the major space agencies, using remote sensing, *in situ* technologies and sample return missions will continue the search for detection of these biosignatures.

9.2.1.4 Chemical History Recorded in Meteorites

Small solar system bodies are believed to be relics of the formation of our planetary system. *In situ* analysis or retrieving a sample from a low-gravity body is significantly different from a sample return from a planet, as has been spectacularly demonstrated by the *Rosetta* mission, projected to orbit, study, and to land on the comet 67P/Churyumov–Gerasimenko (see Section 9.1.2.5). This kind of mission is technically challenging. In order to reach the target, the Rosetta robotic spacecraft travelled a total distance of 6.4 billion km, gaining speed by swinging through the gravitational pull of Earth and Mars along the way, without the possibility to make any physical repairs to the craft during the journey. It arrived at the comet in August 2014 and successfully landed a module on the comet's surface in November 2014. Many space missions have been sent to comets and asteroids over the past few decades. But these are expensive, and only a few have successfully brought back samples. However, there is another possible way to obtain information from Solar System objects. Interplanetary samples have been constantly delivered to Earth over its entire existence. It has been estimated that currently our planet receives about 17 000 meteorites per year, with fragments weighing between 50 g and 10 kg (the latter is very rare). Most of what we know about the Solar System's history comes from these space rocks.

As we have seen in Section 9.1.2, meteorites are ancient objects originating in the asteroid belt that have recorded a succession of chemical processes, starting from reactions occurring either in the interstellar or circumstellar medium followed by reactions that accompanied the formation and evolution of the early solar system, and culminating with reactions during aqueous alteration in their parent bodies. Some meteorites may have originated in comets, but this is highly controversial. These samples may represent only part of the whole picture. In fact, while asteroids are rockier and drier, because they were formed in the inner Solar System, comets formed beyond the snow line, where ices such as frozen water, methane or carbon dioxide can remain stable, giving them, very roughly, a "dirty snowball" composition.

Meteorites are grouped into three major classes: irons, stones and stony-irons. Based on their compositions, stony meteorites are subdivided into chondrites and achondrites. Within the chondrites, there exists a carbonaceous subclass that contains up to 5% in weight of organic carbon. The chemistry and mineralogy of carbonaceous chondrites suggest that this organic matter should be ancient, predating the origin of life on Earth. These materials represent a perfect example of abiotic chemical evolution. Therefore, it can be reasonably assumed that these objects have delivered a significant amount of organic material to early Earth. Such material may have played a major role in the origin of life, perhaps as the original feedstock of organic compounds and/or as catalysts. It is therefore highly relevant to understand the organic composition of these meteorites.

As we have seen in Section 9.1.3, more than 80% of the total amount of the organic carbon in meteorites is a high molecular weight macromolecular material, insoluble in common solvents. This substance is isolated from the bulk rock through successive water and solvent extractions as well as hydrolyses to remove soluble and hydrolyzed organic compounds, and further acid treatments to dissolve most of the mineral matrix. Various analytical techniques including nuclear magnetic resonance, gas chromatography and mass spectroscopy have revealed that more than 60% of the insoluble organic carbon is bound in aromatic components that may be highly substituted. There is also a relatively low bulk hydrogen content and highly branched aliphatic carbon chains.

The soluble fraction of the organic matter in carbonaceous chondrites can be accessed by extracting powdered meteorite samples with solvents of varying polarity. These extracts contain complex mixtures of organic compounds, whose identifications are based on chromatography combined with techniques that detect the eluting compounds. The presence of amino acids, primarily glycine, in meteorites has been known since 1971 when they were detected in the Murchison meteorite. In the decades since, other amino acids and chemical precursors to life have been uncovered in other space rocks. Recent discoveries include nucleobases and sugars, both of them key components of DNA and RNA, and also (a controversial and as yet unpublished result) a protein containing iron and lithium. In general, the soluble organic component is highly complex, with tens of thousands of unique molecular compositions and likely millions of distinct chemical structures, the vast majority of which have not yet been identified. To date, around 100 different amino acids have been identified in the Murchison meteorite, and there are possibly hundreds more amino acids detected that have yet to be explicitly identified and quantified. The list of species unambiguously identified in meteorites includes molecules containing from two to nine carbon atoms.

The exact formation pathway of amino acids in meteorites is unknown. All types of amino acids discovered in meteorites correlate well with the presence of liquid water. Mineralogical evidence from silicate material in meteorites provides strong evidence for the presence of liquid water inside their asteroidal parent bodies for at least a few millions of years after accretion. This aqueous alteration appears to have occurred at a temperature of a few tens of Celsius degrees, induced by low-energy impacts and radioactive decay of short-lived isotopes. Under these conditions, the organic chemical reactions occurring in the condensed phase were completely different from those occurring in the interstellar medium, in either the gas-phase or solid-phase (Chapter 8). It has been thought that some amino acids formed *via* reactions of aldehydes and ketones with ammonia and hydrogen cyanide, in the so-called Strecker-cyanohydrin synthesis. In Strecker's original experiment, acetaldehyde (CH_3CHO) was made to react with NH_3 and HCN to produce the amino acid alanine ($NH_2CH_3CHCOOH$). In the first stage, during the reaction of acetaldehyde with ammonia, nitrogen 'exchanges' with oxygen, and the original carbonyl group becomes CH_3CHNH, an imine. The oxygen reacts with two hydrogens of the ammonia to form water:

$$CH_3CHO + NH_3 \rightarrow CH_3CHNH + H_2O$$

The imine reacts with HCN in a process in which HCN gives a proton (H^+) to the group $=NH$ of the imine, leading to the formation of the amino group $-NH_2$. Note that the double bond is now single, as we have two hydrogen atoms bonded. The remaining cyanide group, $-CN$, bonds to the central carbon forming an aminonitrile:

$$CH_3CHNH + HCN \rightarrow CH_3CH(NH_2)CN$$

In the final stage, upon adding water to the aminonitrile, the $-CN$ group is converted to a $-COOH$ group, NH_3 is ejected and the alanine is formed.

Although this is a very reasonable way to synthesize α-amino acids, *i.e.*, simple molecules that are made of a single central carbon atom bound to a primary amine group NH_2 and to a carboxylic group COOH, other more complex types of amino acids cannot follow this route. β-amino acids, where two carbon atoms separate the amino and carboxyl groups, are believed to have formed by Michael addition, one of the most useful processes for the formation of C–C bonds. Like the Strecker-cyanohydrin reaction, Michael addition requires liquid water.

Nucleobases are other important constituents of the terrestrial biochemistry found in meteorites. They come in two varieties: the purines, guanine $(C_5H_5N_5O)$ and adenine $(C_5H_5N_5)$; the pyrimidines, cytosine $(C_4H_5N_3O)$, thymine $(C_5H_6N_2O_2)$ and uracil $(C_4H_4N_2O_2)$. Two out of the five, cytosine and thymine, have not yet been identified in meteorites. These compounds are typically 1–3 orders of magnitude less abundant in carbonaceous chondrites than glycine, the most abundant amino acid. Among them, guanine shows the largest concentration, followed by adenine and uracil. A survey of reaction mechanisms suggests that Fischer–Tropsch synthesis, *i.e.*, CO, H_2 and NH_3 gases reacting in the presence of a catalyst such as aluminium or silicon oxides, is the most promising route for nucleobase formation under the conditions that characterize the early stages in the evolution of planetesimals. The next step in complexity leads to nucleosides, consisting of a molecule of sugar linked to a nitrogen-containing organic ring compound. In our biochemical machinery, the sugar is ribose or deoxyribose, and the ring, one of the five DNA and RNA bases. Ribose, along with several chemically similar sugars, has been identified in samples from two meteorites, one collected from Morocco, the other from Australia, by using gas chromatography and mass spectrometry. Sugars might have been formed through reactions between water and formaldehyde in the meteorites. This result provides the first direct evidence of the formation of ribose in space and its delivery to Earth.

Amino acids are chiral molecules. These species exist in two forms called enantiomers, mirror images of each other. These forms are chemically equivalent left-handed (L) and right-handed (D) mirror-image forms. Synthetic sugars and other chiral molecules made in the laboratory from non-chiral precursors, or any process that does not use asymmetric reagents such as enzymes in biological reactions, tend to comprise equal amounts (racemic mixtures) of D- and L-enantiomers. For reasons that are still largely unknown, in living organisms, biological polymers, such as proteins and nucleic acids, are made uniquely with left-handed amino acids, and right-handed sugars, respectively. Such an exclusive use of one enantiomer is referred to as *homochirality*. Chirality of biopolymers is of paramount importance in the sophisticated mechanisms of molecular recognition necessary to the machineries of life to perform their complex functions. From the simple stereospecific interaction between an enantiomeric substrate and an enzyme or receptor, to the more complex nucleic acid–protein interactions or DNA duplication, all these processes essential to life are based on a precise selection of the spatial disposition of substituent groups around asymmetric carbon atoms. One of the most significant discoveries of relevance to prebiotic chemistry has been the finding of several non-racemic left-handed amino acids in the Murchison and Murray meteorites. These results have been difficult to explain, since abiotic formation by astrochemical processes, particularly those taking place on interstellar dust (see Chapter 8), and in meteoric interiors (*e.g.*, Strecker-cyanohydrin synthesis), does not appear to be capable of producing them with a nonzero enantiomeric excess.

9.2.1.5 Cometary Chemistry

As we have previously noted, meteorites – humorously called the "poor man's space probe" because they deliver materials from outer space to our door for free – reflect to some degree the composition of that part of the solar nebula from which they formed. They don't therefore provide information on the coldest regions of the protosolar disc, the regions where comets formed. Comets were then scattered by the giant outer planets either into the Kuiper belt, a ring of objects that lies outside Pluto's orbit in the ecliptic plane, or the Oort cloud, where they remain unless collisions or other gravitational disturbances scatter them to the inner solar system. Comets contain tracers of all stages of formation, from the interstellar medium to planets to, perhaps, the origin of life, and this occurs because the bodies experience little chemical activity in the solar nebula. Since their sizes are so small, just a few kilometres in diameter, radioactivity heating was moderate. As a consequence, comets never underwent aqueous alterations, although some high-temperature minerals have been detected, proving that some mixing occurred. Comets present very porous structures, and thus very low thermal inertia, which keeps their interiors cold.

Noble gases are important chemical probes because they undergo few chemical interactions. Their isotopic fingerprints reflect the earliest stage at which the elements formed by nucleosynthesis and point to certain stellar origins. Volatile molecules such as CO and diatomic sulfur record the temperatures experienced by a material during its evolution. Isotopologues such as $C^{16}O$, ^{13}CO, and $C^{18}O$ have different sublimation and condensation rates, and this produces isotope fractionation, which in turn provides direct evidence of cold-temperature chemistry. Finally, comparison of the inventory of organic species detected in comets with those observed in the interstellar medium and star-forming regions may indicate how much disc chemistry is inherited from those cold prestellar environments.

Remote sensing of cometary tails and comae started early in modern astronomy with the appearance of Halley's comet in 1910. The first *in situ* exploration was performed by the ESA *Giotto* spacecraft that approached comet 1P/Halley in 1986. The results provided evidence that comets are much more than "dirty snowballs". Missions were generally limited to flybys because of the extreme difficulty in orbital dynamics of spacecrafts, until the ESA's *Rosetta* probe launched in 2004, and reached its target, the comet 67P/Churyumov–Gerasimenko ten years later. The probe carried eleven instruments, and nine more including two mass spectrometers and a drill on the landing unit *Philae*. The mission's major objectives were: characterization of the nucleus, its dynamic properties, surface morphology and composition; studying the chemistry and mineralogy of refractory and volatile materials; their relation; the origin of comets, and relationship between cometary and interstellar material. Radio science instruments probed the interior of the nucleus, finding a rather homogenous structure, but it is still unclear whether ice and dust are intimately mixed or the ice fills the voids between the dust grains. The mineralogical and structural evidence is of highly porous, low-density objects with very low tensile strength; this suggests an aggregation process that proceeds through gravitational and not dynamical interaction, forming structures across a wide variety of scales. The unplanned landing location of the lander *Philae* turned out to be a lucky event, because the site was cleaned of cometary dust, leaving behind the inner, true, comet surface, with plenty of cracks and granular components. The images reported the very building blocks of the comet, millimetre-to-centimetre-sized aggregates reminiscent of pebbles on a beach, very consistent with the idea that comets emerge by gentle gravitational aggregation of materials. This has, of course, significant implications for planetary formation.

Analysis of the gas by the instruments on board the *Rosetta* orbiter and its lander *Philae* detected all the species viewed previously from remote sensing of 67P/Churyumov–Gerasimenko and other comets, plus some others. This chemical inventory is qualitatively and, in many cases, quantitatively consistent with that determined for astronomical ices, as measured by infrared spectroscopy of embedded proto-stars and background stars, confirming that this material is essentially unprocessed interstellar matter. An unexpected detection has been that of abundant O_2, a highly volatile and extremely reactive species. O_2 is not supposed to maintain significant abundances in hydrogen-dominated environments. Although some local chemical explanations supporting the existence of O_2 have been attempted, *e.g.*, dissociation of water, the data point to a primordial origin for O_2, in the gas- or solid-phases. The detection of N_2, together with CO, CH_4, and Ar, confirms that comets condense highly volatile species from their environments, suggesting temperatures below 30 K during formation.

Phosphorus, a key element in all known forms of life, has also been detected. Analysis of the free dust and of the surface of the nucleus of the comet shows the presence of silicates and organic materials. These materials, as well as molecules ejected by the comet, contain species relevant to biochemistry, including methyl isocyanate (CH_3NCO) and glycine (NH_2CH_2COOH), the simplest amino acid. Mass spectroscopy has also revealed the presence of ammonium (NH_4^+) salts of the form $NH_4^+R^-$. These compounds are formed in reactions between ammonia and acids such as hydrogen cyanide (HCN), hydrogen chloride (HCl), and formic acid (HCOOH). Their sublimation temperatures are higher than those of the individual parts, and upon sublimation, they mostly dissociate again into NH_3 and acid. Ammonium salts are involved in the synthesis of amino acids and the nucleobase adenine. The intermediate step in this reaction is cyanamide (CN_2H_2) that then reacts with glycolaldehyde ($HOCH_2CHO$). If these species are relevant to astrobiology, their detections undermine their utility as biosignatures because, *e.g.*, O_2 together with CH_4 or amino acids is now known to exist in the abiotic world of cometary ice.

9.3 Organic Chemistry in Space

The detection of amino acids in the Murchison meteorite (see Section 9.2.1.4) implies that these compounds are directly produced by reactions in space. This conclusion has been confirmed in one of the most iconic experiments in astrochemistry performed in two independent studies, one at a laboratory at NASA Ames Research Center in California by Max Bernstein and collaborators[1] and the other at the Leiden Observatory in The Netherlands by a team[4] led by Guill-ermo Muñoz-Caro. In these experiments, amino acids, the building blocks of proteins, were found after acid hydrolysis treatment in the room-temperature residue of dirty water ices, supposedly match-ing interstellar compositions, processed with ultraviolet radiation. The ices contained a fairly high amount of ammonia, methanol and hydrogen cyanide. In Leiden, Muñoz-Caro and coworkers found 16 amino acids in the ice residue. They repeated the experiments a sec-ond time replacing all of the carbon atoms in the initial ice compo-nents with ^{13}C isotopes. Again, the same 16 amino acids were found, and mass spectroscopy clearly proved that all the carbon atoms of the generated amino acids were ^{13}C isotopes. In a recent experi-ment, another team,[5] using ices of similar composition irradiated with X-rays, observed infrared and mass spectroscopy signatures of the formation of glycine directly in the ice. The apparently straight-forward production of amino acids is a surprising conclusion, since glycine, the simplest amino acid, has been intensively searched for in space without detection. This may be due to the extreme photo-lability of amino acids, *i.e.*, they are easily destroyed by interstellar ultraviolet radiation.

Could amino acids have survived in space until they were incor-porated intact into asteroids, the parent bodies of the meteorites? It would seem much more likely that simple interstellar precursors may have served as starting material for the Strecker-cyanohydrin synthe-sis (see Section 9.2.1.4) and other secondary processing that inevita-bly occurred on the meteorite parent body. Nevertheless, as, *e.g.*, in Chapter 8, processing of interstellar ices generates chemical complex-ity from mixtures of simple species, inducing the formation of several organic species including ketene (H_2CCO), acetaldehyde (CH_3CHO), ethanol (C_2H_5OH), dimethyl ether (CH_3OCH_3), glyoxal ($HCOCOH$), glycolaldehyde ($HOCH_2CHO$), ethene-1,2-diol ($HOCHCHOH$), eth-ylene glycol ($HOCH_2CH_2OH$), methoxy methanol (CH_3OCH_2OH), glyc-erol ($CH_2OHCHOHCH_2OH$), methyl formate (HCO_2CH_3), formamide ($HCONH_2$) and many others. Molecules containing four to five oxygen atoms have also been found. They include sugars possibly relevant to RNA, phospholipids and energy storage. The list could continue, but the key point is that this rather rudimentary tool to increase chemical complexity in space actually works well. Not all pathways are equally efficient, but the broad range of conditions under which organic com-pounds form demonstrates that their synthesis can occur in a wide variety of astronomical regions.

9.3.1 Exogenous and Endogenous Sources of Organic Materials on Early Earth

The ease of formation of organic compounds of biological interest suggests that exogeneous (*i.e.*, external) delivery was significant. Since there was much more debris in the early solar system, there was, perhaps, a million times more matter falling onto early Earth than there is today (currently about 300 tons per day). Alternatively, terrestrial synthesis from endogenous (local) material may have occurred in Earth's primordial atmosphere and oceans as well as submarine hydrothermal vents. Terrestrial sources depend significantly on the oxidation state of the early atmosphere and the ocean. Such behaviour depends on the efficiency of organic chemistry in reducing or oxidizing environments. Ions or molecules that accept electrons are called oxidizing agents (by accepting electrons they oxidize other species). If they donate electrons they are called reducing agents (by giving electrons they reduce the other species). The ground-breaking experiments performed in 1952 by Miller[6] (under the supervision of Urey) showed that compounds of biochemical importance could be produced in high yields from a mixture of reducing gases.

In those years (the 1950s), it was generally accepted that for early Earth, there was an anoxic (without free oxygen) atmosphere rich in N_2 and other components such as methane, water vapour, and possibly ammonia. A reducing environment such as that would tend to donate electrons to the atmosphere, leading to reactions that form more complex molecules from simpler ones. By contrast, today's oxidizing atmosphere works in the opposite way, stripping electrons from chemical bonds, so that prebiotic molecules would be destroyed as fast as they could be produced. It would be very difficult to produce such molecules in the presence of an oxidizing atmosphere. The reducing conditions in the atmosphere of early Earth as well as high energy levels from ultraviolet radiation and strong lightning and volcanic discharges were thus believed to have set the stage for the origin of life. In 1957, Miller[7] showed that formaldehyde and hydrogen cyanide were key intermediates in the synthesis of glycine. This led Joan Oró and his co-workers[8] to discover, a few years later, that a solution of ammonium cyanide (NH_4CN) in water produces adenine, one of the four bases of DNA. These and other similar discoveries determined the direction of research on prebiotic chemistry for some years. From the late 1960s onward, however, it became clear that the young Earth's atmospheric composition would have been rich in CO_2 (see Section 9.1.4), weakening dramatically the possibility of an abiotic synthesis of organic molecules from endogenous materials in the atmosphere of early Earth. The importance of the oxidation state of the atmosphere is linked to the production of HCN, which seems to be essential for the synthesis of amino acids and purine nucleobases, as well as cyanoacetylene for pyrimidine nucleobase synthesis.

Since their discovery[9] by Robert Ballard in 1977, deep sea hydrothermal vents have been suggested as the birthplace of life. Deep under Earth's seas, seawater comes into contact with minerals from the planet's crust, reacting to create a warm, alkaline (high on the pH scale) environment containing hydrogen. In the process, mineral chimneys are created providing a source of energy that facilitates chemical reactions between hydrogen and carbon dioxide to form increasingly complex organic compounds. In fact, some of the world's oldest fossils originated in such underwater vents. Hydrothermal vents are considered extreme habitats for life, and their very existence shows that life could thrive independent of sunlight. More interesting for astrobiology, they contain elements and conditions conducive to metabolic pathways that some scientists believe were necessary for the evolution of life. The first life must be able to generate the compounds essential for life (metabolism) with its integrity being protected from the environment inside a "container", a structure similar to a cell membrane. The membranes in the cells of every lifeform on Earth all function in about the same way, which suggests that Nature has been using the same basic design for membrane construction since life first evolved. In a recent experiment,[10] it was found that vesicles formed spontaneously under chemical conditions similar to those present in hydrothermal vents, and at pH levels of up to nine (sea water in the open ocean has a pH of about eight). Whether or not hydrothermal vents are the places where life evolved on Earth, similar locations may occur elsewhere in the solar system. Jupiter's moon Europa and Saturn's moon Enceladus are candidates because they both have oceans beneath icy shells (see Section 9.2.1.3).

An estimate of the contribution of different sources of complex organics on early Earth is shown in Figure 9.3.[11,12] Different sources provide their own characteristic organic syntheses, thus contributing different kinds of compounds. However, specific classes of prebiotic molecules, such as, *e.g.*, amino acids, are common outcomes from divergent settings like meteorites, spark discharges and hydrothermal vents.

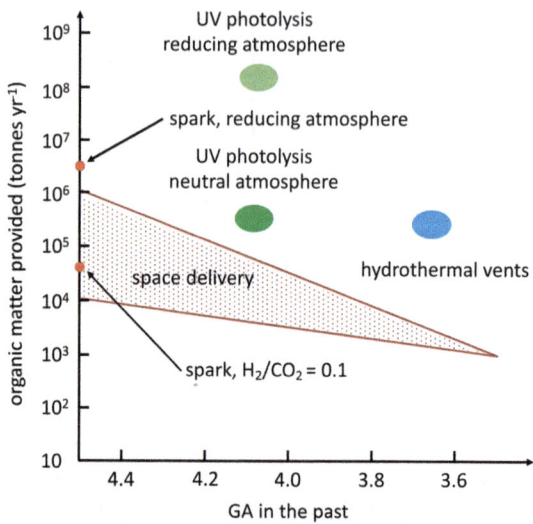

Figure 9.3 Estimated inputs of organic materials from various endogenous and exogeneous sources. The basic graph layout is from a figure by S. L. Miller, as reported in ref. 10. The shaded area represents the uncertainty of the fall-off of dust input on early Earth, estimated by Chyba and Sagan.[12]

9.4   Protected Environments in Space

Early Earth was subjected to the constant threat of impacts with debris left over from the planet-building phase. This violent activity occurred from 4.1 to 3.8 billion years ago. Close to the end of this period, impacts in the Solar System may have increased, because of the migration of giant planets, which sent debris into scattered inner orbits that intercepted those of smaller rocky worlds. The evidence for two early-bombardment populations is now emerging, and that there was a time interval between them. The later bombardment came plausibly from the asteroid belt, while the early one came from elsewhere. The first episode might have been caused by failed young planets or planetesimals – thought to be far larger than the objects in the asteroid belt – that would have done significant damage as they crashed into the rocky inner solar system planets. During this dramatic era, it is thought that most of Earth's water would have been vaporized, and perhaps the exterior of our planet would have been sterilized, erasing any life forms that might have succeeded in emerging. If so, pre-biotic chemistry had to wait for the storm to pass before it could take firm root and eventually give rise to all later life. The oldest fossils are at least 3.8 billion and possibly 4.3 billion years old, suggesting an "almost instantaneous emergence of life" (at least in the geological sense), after ocean formation occurred 4.4 billion years ago. Earth is estimated to be about 4.5 billion years old, and thus for much of its history it has been home to life. Was biological evolution on Earth jump-started by a special delivery from outer space?

The importance of exogenous delivery of organic matter to early Earth is critically dependent on the survivability of organic compounds during the formation and delivery stages. They should evolve in a protected environment able to shield them from the extremely unfavourable environmental conditions at the beginning of the Sun's life. The physical conditions and chemical reactions of the kind Stanley Miller and others envisaged and explored in early pre-biotic experiments may not be characteristic only of planetary environments. We have seen that circumstellar discs arise naturally during the formation of a low-mass star. During the evolution of these discs, molecules freeze-out from the gas phase onto dust grain surfaces while dust grains agglomerate. This process ends up in the formation of loosely packed structures with much of their internal volume being vacuum and trapped ices. Interstitial voids must occur, even in highly organized, densely packed structures (see Figure 9.4).

As we saw in Chapters 7 and 8, surface chemistry proceeds in a sequence of steps. First of all, a future reactant must collide with the grain surface, and then accommodate and stick; it may then react at the site of impact, or undergo a delayed reaction with other grain adsorbates. The products of surface reactions are either retained on the dust or desorbed to the ambient gas.

The internal voids produced by the agglomeration process offer a different intermediate possibility: the re-accretion of reaction products onto other components of the aggregate. As desorbed products can be in an energetic state, these secondary reactions might mimic some aspects of high-temperature chemistry. In addition to the chemistry produced by ice processing, additional chemical reactions could occur later on during the aggregation phase. Dust aggregates would be impulsively heated by collisions with other grains and by cosmic ray impacts. During the interaction of high velocity cosmic ray particles with dust, the atoms of the grain materials (C, Si, Ca, Mg, Fe, and P) are dislodged and ejected from the surface. The heat released during a collision may lead to vaporization of the ice content filling the cavities (in a process called sputtering). In both types of impact the sublimated ices would then enter a transient, warm, high-pressure gas phase, together with sputtered atoms from the grain substrate, in a hydrogen-rich atmosphere.

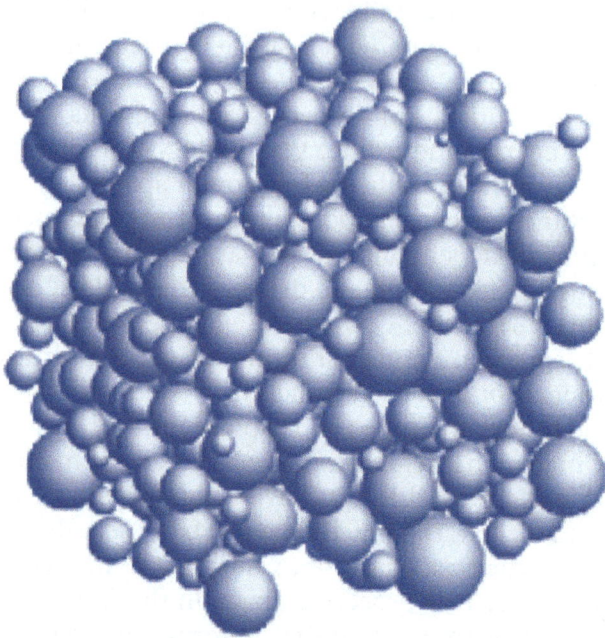

Figure 9.4   A cluster of grains. The image represents a cluster of many
spherical grains with a range of sizes. It indicates that much of
the volume inside such a collection must be available, and may
be occupied either by interstellar gas or vacuum.

This is a unique astrochemical situation. The resulting mixture of
volatile and refractory materials would be subjected to similar conditions to those experienced in the Urey and Miller prebiotic experiment. Grain aggregates represent thus the equivalent of terrestrial
micro-laboratories containing raw materials of reducing chemical
composition suitable for conversion into more complex organic species.[13] In other words, the Urey–Miller experiment may have been
revisited through innumerable repetitions inside dust grain aggregates. The chemical products contained in these aggregates would
then be incorporated into planetesimals and comets that may fall on
a planet, leading to a plausible connection between the chemistry in
cold interstellar regions and that in material falling onto planets. We
don't yet know whether this Fischer–Tropsch type of processing would
favour certain types of product molecules.

As we have seen above, long after the planets of the Solar System
formed, catastrophic collisions continued, with a climax about 4 billion years ago. The scars of this geological violence are evident today
in the ancient cratered terrains of planetary surfaces. These giant

impacts had destructive effects and would have obliterated any form of chemical complexity either in the impactor interiors or in the primordial Earth. However, the gentle dusty rain of clusters of small dust particles provided a continuous supply of organic chemically organized materials. If life did originate on Earth's surface, then it could have happened many times before the end of the major impact era, or, if it started only after the conditions became less hostile, it succeeded very quickly.

9.5 From Selected Molecular Building Blocks to Life Itself

Life's simplest molecular building blocks, including amino acids, sugars, lipids and bases, emerged inexorably through simple, inevitable chemical reactions occurring in a variety of suitably energized prebiotic environments. However, as we have seen in this chapter, abiotic synthesis encourages the formation of molecules made of fewer rather than greater numbers of carbon atoms. When larger carbon-containing molecules are produced, they tend to be insoluble, hydrogen-poor substances. Chemical processes make small molecules, but one characteristic that differentiates life from other phenomena is that life consistently utilizes and generates complex molecules. In other words, life requires the assembly of just the right combination of specific small molecules into much larger collections which have specific functions. Assembling the building blocks into useful macromolecules is complicated by the fact that the vast majority of carbon-based molecules synthesized in prebiotic processes have no obvious biological use. The overarching problem with studying life's origins is the realization that the complex molecules involved are deeply encrypted. DNA in a cell decodes itself with the help of an entire entourage of biochemical helpers. For example, DNA depends on proteins for functioning, and also contains the recipe for protein construction. Which comes first, the DNA or the protein? This is the so-called DNA paradox.

The resolution of the paradox may lie in the "RNA World" hypothesis. In this view, genetic information flows from DNA to RNA to proteins: DNA encodes RNA which directs the synthesis of proteins. Proteins do most of the work in cells, and capture energy used in the synthesis of new copies of DNA, enabling the formation of new copies of DNA and growth of new cells and organisms. Proteins are very efficient in all these tasks because they speed up reactions of biological systems by lowering the activation barrier inhibiting the reaction. The protein's shape, encoded in the RNA, determines its function.

Structurally, DNA and RNA are nearly identical. Both are polymers of monomers called nucleotides, but DNA has two strands (the double helix) while RNA has only one. The RNA World hypothesis proposes that there was a primordial soup which included free nucleotides. The nucleotides formed chains, and evolved self-replicating RNA patterns that enabled a range of catalytic functions. Then RNA began to synthesize proteins which took over some of the RNA functions. Finally, DNA appeared; its main function is as a reliable holder of genetic information.

The RNA world model is not without its difficulties. One of the major problems, which turns out to be a necessary intermediate step in the chain of events described above, is to recreate a scenario in which functional and replicating RNA could have self-assembled. With the current technologies, and a huge array of available chemical building blocks, the complexity of newly synthesized organic molecules is limited only by the imagination of the experimenter. Prebiotic chemistry, on the other hand, is not afforded such luxuries. Nevertheless, we have seen that in meteorites the basic functional units of nucleosides (subunits of nucleotides), bases and sugars, are (almost) routinely found. A further necessary step would be finding a convincing prebiotic synthesis of nucleosides, *i.e.*, a way to assemble the ingredients already present in the environment. However, even if a prebiotic synthetic pathway to nucleotides could be found, we still need to link those nucleotides into an RNA strand, a task that appear to be of exceptional difficulty. Perhaps, some speculate, a simpler nucleic acid preceded RNA. All these chemical steps had been driven by competition. Natural systems proceed through random variations in the chemical makeup of key biomolecules, producing some neutral species, others doomed to failure, and once in a while providing some constituting an improvement. This mechanism has also been the driving force of the evolution of life over the past four billion years of Earth's history.

Appendix: Glossary of Some Technical Terms Used in Astrobiology

Abiogenesis: the concept that the origin of life is a process in which life arose naturally from reactions in non-living matter

Abiotic: not derived from living organisms

Alanine: an amino acid used in the biosynthesis of proteins. It carries a methyl group side chain.

Amino acids: organic compounds containing amino and carboxylate functional groups, along with a side chain specific to each amino acid. They are added step by step to grow a protein chain.

Asteroids: minor planets in the inner solar system

Biogenic: a biogenic substance is a product made by or of life forms

Biomolecule: molecules produced by living organisms

Biosignature: any substance or event that provides evidence of past or present life

Biosphere: the region in which life occurs on Earth

Fischer–Tropsch process: the conversion of carbon monoxide and hydrogen or water gas into liquid hydrocarbons

Glycine: the simplest amino acid. It carries a single H atom as its side chain.

Lipids: macrobiomolecules, including fatty acids, insoluble in water but soluble in organic solvents

Meteoroids: small rocky bodies in space. When a meteoroid enters Earth's atmosphere it creates a flash of light called a *meteor*. If any of the meteoroid survives to land on Earth's surface, the remaining solid is called a *meteorite*. Meteoroids contain organic molecules including amino acids.

Nucleobases: nitrogen-containing biological molecules that form *nucleosides* which are components of *nucleotides*, the building blocks of RNA and DNA

Planetesimals: solid bodies in protoplanetary discs formed from the accumulation of dust grains. They are gravitationally bound objects and are the precursors of planets

Prebiotic chemistry: inorganic or organic chemistry of natural materials before the advent of life

Presolar dust: cosmic dust found in the solar system but formed long before the Sun appeared

Prestellar core: an early phase of the formation of low mass stars, before gravitational contraction creates a protostar

Proteins: large biomolecules capable of performing various tasks, including catalysis of reactions, DNA replication, responding to stimuli, transporting molecules. Proteins are defined by their sequence of amino acids; this determines their folding into structures that determine their activity

Protostar: a very young star still accreting mass from its parent cloud

Ribose: a simple sugar, a component of ribonucleotides from which RNA is built. Its analog, deoxyribose, is an essential component of DNA

RNA and DNA: ribonucleic and deoxyribonucleic acids; these are polymeric molecules in the form of a chain of nucleotides, and are essential for all known forms of life

Replication: DNA replication is the process in which two identical copies are made of the original DNA molecule

Serine: an amino acid used in the biosynthesis of proteins. It carries a hydroxymethyl group side chain.

Strecker synthesis: a method of forming amino acids by reacting an aldehyde with ammonia in the presence of potassium cyanide

Sugars: carbohydrates of general formula $C_n(H_2O)_n$ including glucose and fructose or the disaccharide sucrose which contains one molecule of each

Translation: the process of translating the sequence of a messenger RNA molecule to a sequence of amino acids during protein synthesis

References

1. M. P. Bernstein, J. P. Dworkin, S. A. Sandford, G. W. Cooper and L. J. Allamandola, *Nature*, 2002, **416**, 401.
2. P. Nurse, *What Is Life?*, David Fickling Books, Oxford, 2020, ch. 4.
3. https://exoplanets.nasa.gov.
4. G. M. Muñoz Caro, U. J. Meierhenrich, W. A. Schutte, B. Barbier, A. Arcones Segovia, H. Rosenbauer, W. H.-P. Thiemann, A. Brack and J. M. Greenberg, *Nature*, 2002, **416**, 403.
5. A. Ciaravella, A. Jiménez-Escobar, C. Cecchi-Pestellini, C. H. Huang, N. E. Sie, G. M. Muñoz Caro and Y. J. Chen, *Astrophys. J.*, 2019, **879**, 21.
6. S. L. Miller, *Science*, 1953, **117**, 528.
7. S. L. Miller, *Biochim. Biophys. Acta*, 1957, **23**, 480.
8. J. Orò and A. Kimball, *Biochem. Biophys. Res. Commun.*, 1960, **2**, 407.
9. R. D. Ballard and Tj. H. Van Andel, *Bull. Geol. Soc. Am.*, 1977, **88**, 507.
10. S. F. Jordan, H. Rammu, I. N. Zheludev, A. M. Hartley, A. Maréchal and N. Lane, *Nat. Ecol. Evol.*, 2019, 3(12), 1705.
11. M. Bernstein, *Philos. Trans. R. Soc., B*, 2006, **361**, 1689.
12. C. Chyba and C. Sagan, *Nature*, 1992, **335**, 125.
13. W. W. Duley, *Mon. Not. R. Astron. Soc.*, 2000, **319**, 791.

10 Conclusions

"The last 13 billion years in the history of the Universe have been characterized by a persistent and inevitable increase in chemical complexity." (Cesare Cecchi-Pestellini, Chapter 9 of this book)

10.1 The Surprising Milky Way

Perhaps the most immediate conclusion that a chemist might draw from the discoveries of astrochemistry is one of *surprise* that chemistry can occur in great variety, even in gas at very low temperature and at a pressure lower even than that obtainable in ultra-high vacuum apparatus in chemical laboratories. Of course, this chemistry must be stimulated by electromagnetic or particle radiation, or by local heating, and the interstellar gas must contain reactive atomic species and must be given enough time for chemistry to proceed, for reaction rates in interstellar gas are very slow by terrestrial standards. As we have seen in this book, these requirements can be met in the interstellar medium. But it still seems remarkable that (to take one common astronomical situation) interstellar molecular clouds with, say, a few thousand H-atoms per cm^3, and about one C atom, one O atom, and one N atom per cm^3 (together with some other species of even lower abundance) can generate a wide variety of interstellar molecules in such abundances that they are very clearly detectable by astronomers and have fundamentally important consequences for astronomy in terms of star and planet formation, and – possibly – even astrobiology.

Astrochemistry: Chemistry in Interstellar and Circumstellar Space
By David A. Williams and Cesare Cecchi-Pestellini
© David A. Williams and Cesare Cecchi-Pestellini 2023
Published by the Royal Society of Chemistry, www.rsc.org

This cosmic chemistry takes place in an environment which appears at first sight to be quite unsuitable: interstellar space is flooded by ultraviolet starlight from massive stars; this ultraviolet radiation is capable of dissociating and ionizing many molecular species that might possibly be formed in interstellar chemistry. Interstellar space is also swept by a flux of high energy particles (mainly protons) ranging in energy from about an MeV to about 10^{20} eV *per particle*, and collisions of such particles with molecules are destructive; and interstellar gas is repeatedly hit by high velocity shocks that raise the gas temperature to high values sufficient to dissociate molecules and ionize atoms. This is ostensibly a hostile environment for chemistry, and the famous astronomer Sir Arthur Eddington was right to be sceptical in 1927 about the existence of interstellar molecules (as we noted in Chapter 1). Circumstellar regions are denser and warmer than interstellar ones, yet these regions too can be subject to intense stellar radiation fields and to gas dynamical shocks. Yet we have discovered in this book that interstellar and circumstellar molecules *can* be formed at sufficiently high rates to overcome these losses and accumulate detectable abundances. Chemists *should* be surprised at the evident success of chemistry under these challenging astronomical conditions.

In the preceding chapters in this book we have discussed the ways in which chemistry in interstellar and circumstellar space can begin. We've seen (in Chapters 3 and 4) how a very few highly specific entry routes involving simple chemical reactions in interstellar clouds can generate products that are then able to take part in a wide range of further reactions. Thus, a small and very restricted set of highly specific initiating reactions easily generates enormous networks of very efficient gas phase reactions. We have explored (in Chapter 5) how we can use libraries of these reactions (with measured, computed, or estimated rate coefficients) and efficient computer software to explore the chemistry arising from these networks. From several examples of such computations it is clear that gas phase reactions in denser regions of interstellar space generate a wide variety of interstellar molecules with a remarkable range of complexity.

We have seen that gas phase chemistry in the atmospheres and envelopes of some particular types of cool star can lead to the formation of dust grains (Chapter 6) and that this dust is ejected into interstellar space where it can be chemically active in interstellar clouds (Chapter 7). Surface reactions generate the seminal molecule molecular hydrogen (involved in the gas phase formation of almost all interstellar molecular species) and other species, while the accumulation of simple molecules from the gas phase onto the surfaces of cold dust grains in denser interstellar clouds may form solid ices (Chapter 8). Solid–state chemistry in these ices, stimulated by fast particles or by energetic radiation, creates molecules of relatively high complexity, some of which may even be the feedstock for the formation of much larger, much more complex biological molecules in various scenarios arising during planet formation (Chapter 9).

The highly specific nature of the essential initial chemical steps in all these chemistries might suggest that it is very difficult for inter-stellar chemistry to develop in interstellar clouds. Similarly, the very restricted nature of the circumstellar locations of dust formation might lead to the conclusion that it is difficult to form dust in cir-cumstellar gas. These impressions would be incorrect. What we learn from observations of molecules and dust in the Milky Way is entirely the opposite.

Molecules and dust are widespread in the Milky Way: wherever molecules and dust *can* form in the Milky Way, they *do* form.

In fact, as chemists, we could regard the Milky Way galaxy essentially as *an efficient machine for making molecules and dust.* Wherever the physical conditions in this machine are suitable, molecules and dust are produced. As astrochemists, it is our job to understand how this machine works, to identify the machine's controlling levers, and to discover the consequences for the Milky Way of the presence of molecules and dust. Working towards these aims has been our purpose in this book. Although we have chosen to introduce astrochemistry by referring specifically to conditions in the Milky Way galaxy, this is not a restriction. As we have seen, the Milky Way is merely one of very many galaxies that may differ in evolutionary age and (consequently) in physical conditions, but all the processes we have described here are general and apply to a greater to lesser extent in all galaxies.

As we emphasized in Chapter 1, most of the *volume* of the Milky Way is occupied by very hot, very low density gas in which mole-cules and dust do not exist. Almost all of the *mass* of the interstel-lar gas in the Milky Way, about 10% of the mass of this galaxy, or $\sim 10^{11}$ $M_{\odot}$, occupies a very tiny fraction of interstellar space; this gas is mostly very cold and relatively dense compared to the hot component, and is almost entirely molecular. The total dust mass in the Milky Way is only about one percent of the interstellar mass. The dust is generally well mixed with the gas. The interstellar gas is important because it is the reservoir of matter from which new stars and planets will form. Galaxies which have a significant amount of interstellar matter, like the Milky Way, are making new stars; the Milky Way makes (on average) about one new star every year. Other galaxies may be more active or less active than the Milky Way, depending on the amount of interstellar matter they contain. Molecules and dust, therefore, are tracers of this reservoir of matter. Molecular emissions and absorptions, and dust emissions and extinctions, are the ways that astronomers can trace this gas in any galaxy.

10.2   Astrochemistry and Chemistry

The demands made by astronomers for chemical data of reactions in large astrochemical networks have stimulated remarkable activities in chemical laboratories and in theoretical chemistry. As we have seen in the preceding chapters, these chemical activities and their conclusions include the following:

- Determination of thousands of measured, computed, or estimated values of rate coefficients for relevant reactions to be found in the libraries of reactions described in Chapter 5. The provision of these data is an ongoing task since the libraries of reactions are continually revised and extended;
- Determination of cross sections for the excitation and ionization of many relevant astronomical molecules by electromagnetic radiation over a wide range of wavelengths (especially including the infrared, visible and ultraviolet);
- Evaluation of the response of relevant astronomical molecules to interaction with fast particles (cosmic rays, mainly protons and helium nuclei), especially those with particle energies peaking near MeV;
- Laboratory study of the optical properties of chemically different solid materials, either singly or mixed, to support investigations of the chemical and physical nature of interstellar dust grains;
- Both laboratory and theoretical studies of the interaction of hydrogen atoms with the surfaces of materials plausibly representing interstellar grains, and of the interaction of those hydrogen atoms with each other to form molecular hydrogen, have been carried out so that we can infer the probability of molecular hydrogen formation on interstellar dust, its retention or ejection and – if ejection is prompt – the molecule's kinetic and rovibrational energies;
- Laboratory studies of mixed ices composed of simple molecular species on low temperature surfaces have been performed, so that we may infer the range of physical environments in which molecules are confined in ices formed on interstellar dust;

- Laboratory studies of chemical processing of simple low-temperature mixed ices induced by electromagnetic or particle radiation of various kinds have been carried out so that we may explore the chemical complexity produced during this processing and compare the laboratory complexity with that of the observationally detected interstellar and circumstellar species;
- Theoretical studies of the survival and transfer of interstellar and circumstellar molecules (some of which may be prebiotic species) into regions of star and planet formation have begun, and the roles that these molecules play in interplanetary dust, comets, asteroids and proto-planets are being explored.

Evidently, chemistry has contributed and continues to contribute very successfully to astronomy through its contributions to astrochemistry, in very many essential and fundamental ways. The contributions of chemistry to astrochemistry enable striking advances to be made in astronomy.

10.3   Astrochemistry, Astronomy and Astronomical Observatories

Given this very extensive support from fundamental chemistry, astrochemistry has been able to make enormous contributions to astronomy. Since astronomy is an observationally-driven science, these advances are strongly linked to advances in observational techniques, especially to the use of new facilities that have opened new wavebands for astronomical observation. Until the mid-20th century, only the visible waveband was available, but eventually facilities in the radio, millimetre, infrared, ultraviolet, and X-ray wavebands were introduced as new technologies became available. Ground-based or orbiting observatories were developed, as appropriate, depending on Earth's atmospheric transparency in the waveband selected. Figure 10.1 shows that the atmosphere is effectively opaque for almost all ultraviolet wavelengths but has low opacity in the optical "window" (which is somewhat wider than the range for which the human eye is sensitive, about 400–700 nm). The opacity is highly variable for much of the infrared and millimetre wavebands. There is a radio "window" of low opacity at wavelengths from about a centimetre to about ten metres. In the near infrared and millimetre wavebands, transparency occurs in a number of narrow bands. To operate astronomical instruments in wavebands outside the optical and some selected bands in the near

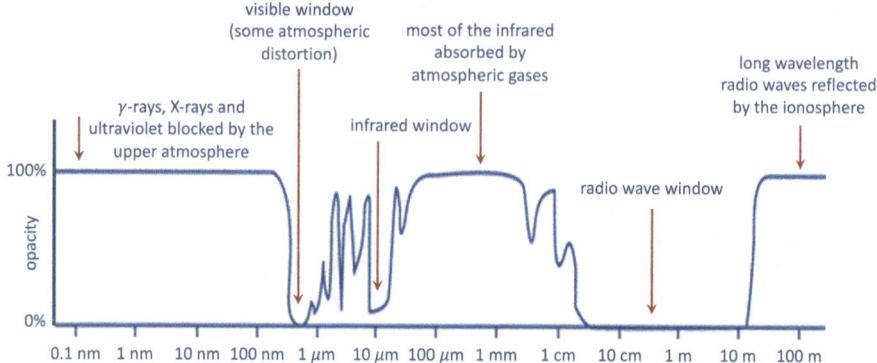

Figure 10.1 Transparency of Earth's atmosphere. Transmission of electro-magnetic radiation through Earth's atmosphere is shown in the diagram to be impossible for very short and short wavelengths (gamma rays and ultraviolet) where opacity arises mainly from absorption of incoming radiation by oxygen and ozone mole-cules. The optical window of low opacity opens at wavelengths near 300 nm (near ultraviolet) and extends until about 700 nm where the near infrared begins. Transmission in the near- and mid infrared is strongly wavelength-dependent, but several narrow windows occur in which astronomical ground-based observations are possible. Observations in the sub-millimetre and millimetre regions are also possible in several wavebands. Opacity in the infrared and millimetre wavebands is mainly due to absorption by water molecules (so astronomical observato-ries are often placed at high altitude, above most of the cloud layers) and also by carbon dioxide molecules. The radio win-dow has near zero opacity and extends for wavelengths from about a centimetre to more than ten metres. For longer radio wavelengths the atmosphere is opaque, the absorption at long radio wavelengths arising in the ionized plasma present at high altitudes. Ground-based astronomical facilities are useful only in the optical window, partially in the infrared and sub-millime-tre/millimetre wavebands, and in the radio window.

infrared, submillimetre and millimetre, observatories must be placed in Earth's orbit.

The vast majority of discoveries of the molecular species listed in the Appendix to Chapter 1 were made by observations of their rota-tional transitions at millimetre and sub-millimetre wavelengths. About fifty ground-based telescopes in many countries operating in these wavebands have contributed to these discoveries, but – of this large number – several telescopes have been pre-eminent. These include the IRAM 30 metre (operated by the Institut de Radio-astronomie Millimétrique, Sierra Nevada, Spain, at an altitude of 2850 m) and the (US) National Radio Astronomy Observatory

(NRAO) 36 foot telescope, renovated in 1984 and renamed the NRAO 12 metre (now operated by the Arizona Radio Observatory at the Kitt Peak National Observatory, USA, at an altitude of 1914 m). These two telescopes have been remarkably successful and between them are credited with many of the astronomical molecular detections listed in the Appendix to Chapter 1 and with detailed studies of the regions in which the molecular species are found. The telescopes are single-dish. An image of the IRAM telescope is shown in Figure 10.2.

The situation in millimetre and submillimetre wave astronomy has been transformed by the Atacama Large Millimetre Array (ALMA), the largest and most sensitive instrument at millimetre and submillimetre wavelengths. It is a multinational project, and has been fully operational since 2013. ALMA is an astronomical interferometer of 66 high-precision linked radio telescopes, of which there are fifty 12 metre antennas in the so-called main array, with sixteen antennas (four of 12 metres and twelve of 7 metres) comprising the compact array. It is sited on the Chajnantor Plateau in the Atacama Desert, Chile, at an altitude of 5000 m. The antennas can be moved over the desert for distances up to 16 km, giving ALMA exceptional angular resolution (~10 milliarcseconds). ALMA operates in all atmospheric windows between 350 μm and 10 mm. ALMA has been used to map

Figure 10.2 An image of the IRAM 30 metre telescope The panels comprising the IRAM 30 m antenna are adjusted to maintain the desired parabolic form during tracking of an astronomical source, to a very high accuracy. The telescope is equipped with heterodyne receivers and continuum cameras operating at wavelengths around 3, 2, 1, and 0.8 millimetres. Reproduced under Creative Commons Attribution-Share Alike 4.0 International license.

Figure 10.3   The Atacama Large Millimetre Array (ALMA). Part of the ALMA Compact Array (ACA). Reproduced under Creative Commons Attribution-Share Alike 4.0 International license.

molecular emissions in galaxies, molecular clouds, star-forming regions, and comets. It has mapped dust emission in planet-forming regions. An image of part of the ALMA array is shown in Figure 10.3.

10.4   Molecules and Dust: Tracers of the Evolution of Interstellar Structures, and Signposts to the Abiogenetic Origin of Life

Astronomers regard molecules and dust grains in astronomy primarily as tracers of matter. The direct observation of molecular lines gives information about the density and temperature of the location in which those molecules are detected and emitted radiation from dust is also an important tracer of these physical parameters. When the chemistry and physics of these molecules are explored using models such as those described in Chapter 5, the range of information obtained from those observations is both more precise and much broader, so information about the local physical conditions (*e.g.*, radiation intensity, fast particle flux, *etc.*) also becomes available. The essential point is that observations of astronomical molecules give a detailed account of the location and condition of those molecules. From this information we are able to infer how this material will evolve.

Observations of molecules in interstellar space therefore trace the various structures of interstellar gas. As we have seen, the range of structures identified begins with diffuse clouds and includes dark clouds, giant molecular clouds, infrared dark clouds, filamentary structures, hot cores, and star- and planet-forming regions. This sequence of structures suggests an evolutionary sequence and is the basis for modern ideas about the formation of stars and planets, as illustrated in the sketch in Figure 1.6. Every stage in this sequence is traced and probed by emissions from either molecules and/or dust grains. Figure 10.4 shows astronomical images of several key stages in this evolution. The sequence is driven by gravity and by gas dynamical flows initiated by supernovae explosions that occur near the end of the lives of massive stars, and by stellar winds from cool stars of several solar masses. Both of these locations are also situations in which astrochemistry plays an important role in generating new gas and dust populations and ejecting them into space.

The discussion of star and planet formation in Chapter 9 identified several situations in which – as experiments confirm – complex chemistry in meteorites leads to the formation of organic species such as amino acids, nucleobases, and sugars. These molecular species are important sub-components of the biological polymers RNA and DNA. Is it possible that life may originate spontaneously from these materials, under suitably favourable conditions? We do not know, but observational evidence and experimental results seem to support the concept of abiogenesis.

10.5 Those Questions in Chapter 1

With the insight gained from reading this book, we can now try to address those questions that we posed in Chapter 1 (Section 1.3). Can we provide convincing answers? Is there still a lot of work to do? What are the areas where new work is needed?

1. Galaxies are sources of powerful particle and electromagnetic radiations that are hostile to molecules, so can we devise chemical networks that successfully overcome the consequent losses and produce species in the abundances observed?

    Yes, we can. These model chemistries are explored in Chapter 5. The most energetic electromagnetic radiations from hot stars in the far ultraviolet are trapped in regions around those stars (the so-called HII regions). However, radiation with wavelengths longer than 91 nm escapes from HII regions and can certainly interact with chemistry in interstellar clouds, destroying molecules. What allows chemistry to proceed in interstellar clouds is the extinction caused by interstellar dust which scatters

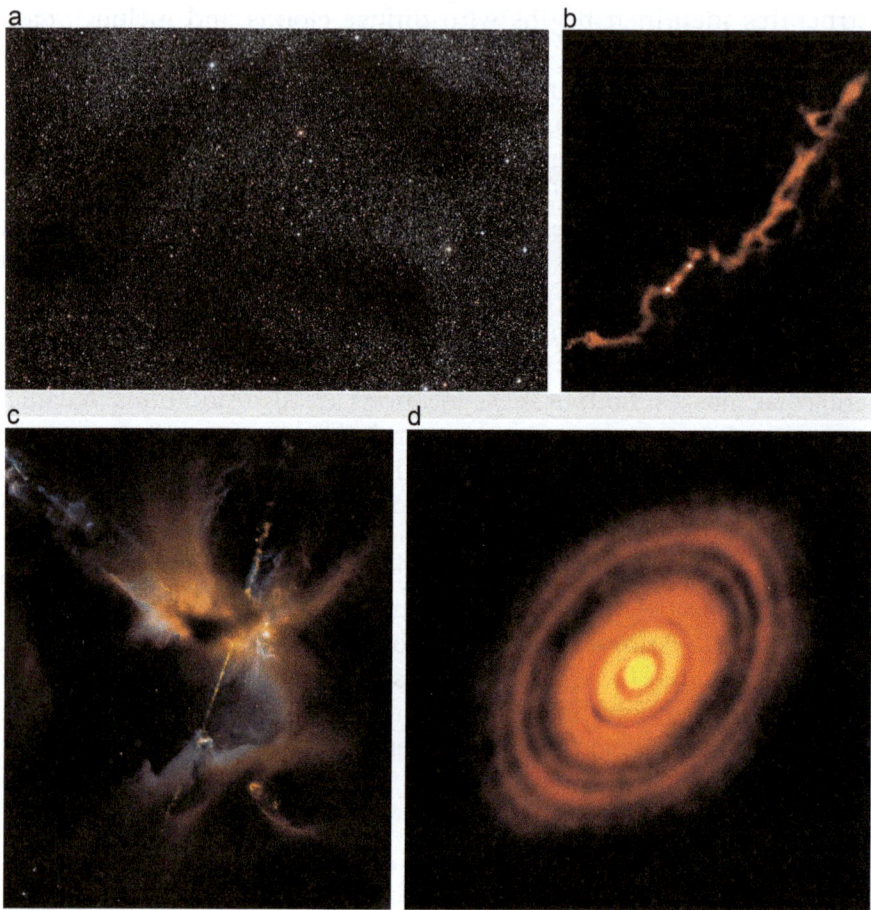

Figure 10.4   Images illustrating some stages in the sequence of events in the formation of stars and planets. (a) An optical image of the dark cloud LDN-1768. Credit: ESO. (b) An infrared image of the dark cloud TMC-1 showing that star formation is occurring on a filament of infrared-emitting warm gas within the cloud. Credit: ESO/APEX (MPIfR/ESO/OSO)/A. Hacar *et al.*/Digitized Sky Survey 2. Acknowledgment: Davide De Martin. (c) A newly formed star is still hidden within a dense core. The star is generating these very striking jets which are part of the stellar outflow that will eventually remove the core, leaving behind a protoplanetary disc around the central star. Credit: NASA, ESA, Hubble Heritage (STScI/AURA)/Hubble-Europe Collaboration; D Padgett (GSFC), T Megeath (University of Toledo), B Reipurth (University of Hawaii). (d) The protoplanetary disc surrounding the young star HL Tauri. This false colour infrared image from ALMA shows the emission from warm grains in the disc. The dark bands may suggest that a number of planets are forming and orbiting the central star. The image is courtesy of ALMA (ESO/NAOJ/NRAO).

and absorbs starlight. Dust extinction can reduce the intensity of stellar ultraviolet inside a dense cloud to a negligible amount.

However, dust does not impede the penetration of cosmic rays (energetic particles) into interstellar clouds. They cause ionizations of atoms and molecules, encouraging fast ion–molecule reactions to occur. Thus, cosmic rays stimulate interstellar chemistry – if the flux of these particles is not too high. In the Milky Way, the flux is fairly low, so cosmic rays have a largely beneficial effect on chemistry. However, this may not be true in all other galaxies, or in particularly active locations in the Milky Way.

2. Much of interstellar gas is at very low densities and temperatures and chemistry is likely to be slow, so is there enough time available for molecular abundances to grow?

   Models such as those developed in Chapter 5 show that the chemistry in a static interstellar cloud under typical steady conditions may take a million years or so to achieve chemical steady-state. Many clouds do appear to be long-lived and in a chemical quasi-steady-state. However, in some situations, physical conditions may change on short timescales, and in those cases the chemistry may never achieve steady-state but is continuously evolving. For example, in a gravitationally collapsing cloud leading ultimately to star formation, gas may begin its accelerating collapse at low density and cool temperatures, but is continuously compressed by gravity and warmed by radiation from the newly-forming star. Under these changing physical conditions, the chemistry never achieves steady-state but must be followed in time throughout its evolution.

3. The interstellar gas is H-rich and other elements are merely present in trace values, so why are some (but not all) of the observed species H-poor (*i.e.*, chemically unsaturated)?

   Insertion of hydrogen from H_2 directly into a carbon chain (for example) is known as hydrogenation and is a slow process in interstellar clouds, while even in the laboratory or in industry hydrogenation requires a metal catalyst and a high temperature. Hydrogen from H_2 can be inserted into a molecular ion (*e.g.*, HCN^+) to form a new ion (*e.g.*, $HCNH^+$); in the interstellar medium this new ion will lose its additional hydrogen in fast dissociative reactions with electrons or molecular cations, or in reaction with neutral molecules. Detailed models such as those described in Chapter 5 show that the low temperature addition of H atoms from H_2 to unsaturated interstellar species does not always proceed readily.

4. The list (see the Appendix to Chapter 1) of detected species includes relatively large numbers of diatomic and triatomic species, and ends (with few exceptions) with relatively small numbers of molecular species

containing about a dozen atoms. Is this a real chemical effect or is it associated with observational difficulties of detecting the larger species? Does the number and variety of interstellar species really decline with complexity?

The decline in species abundance with size may be a real effect rather than due to observational difficulties (although these, too, may present a valid difficulty in the detection of large species). To grow larger and larger species requires a steady input of smaller species. Eventually, this input cannot be maintained and the synthesis of the larger species ceases.

Nevertheless, we showed in Chapter 2 that large molecules have small rotational constants and therefore large molecules have closely spaced rotational levels, implying that the number of rotational levels accessible at any temperature is large. Therefore, the number of molecules in any particular rotational level is small and the emission intensity in any particular transition is weak, and detection becomes more difficult.

5. Is the chemistry producing the fullerenes – and possibly the PAHs – simply an extension of the chemistry producing the simpler molecules listed in the Appendix to Chapter 1, or are different processes likely to be required to form these more complex molecules?

Gas phase schemes for the production of large carbon species such as fullerenes and PAHs have been proposed. These schemes depend on the same types of gas phase reactions that are used to account for the observations of simpler molecules listed in the Appendix to Chapter 1. These schemes have had considerable success. An alternative mode of forming both fullerenes and PAHs would be from the degradation of solid carbon dust grains in interstellar shocks, producing smaller and smaller units, eventually arriving at molecular structures. This alternative has not been fully explored.

6. Silicon and iron have quite similar relative elemental abundances in the interstellar medium of the Milky Way, so why does Si appear in thirteen detected species while Fe appears in only one?

The elemental abundances in the interstellar medium relative to 10^6 H-atoms are 31.6 and 33.1 for Si and Fe, respectively, as measured in the atmospheres of hot stars, so the relative elemental abundances are indeed similar (see Table 1.4). However, the amount of Fe locked in interstellar dust is 32.8 while that for Si is 29.4 (in the same units), so there is slightly more Fe than Si in the dust. The residual amounts remaining in the gas are 2.2 for Si and 0.3 for Fe, so there is seven times as much silicon in the gas as iron. This discrepancy strongly favours the detection of silicon species. Also, silicon has a rich chemistry, similar to that of carbon chemistry (if not as extensive), while iron chemistry is much more limited. Therefore, interstellar molecules bearing silicon would be expected to be more evident than those of iron.

7. Why are some detected interstellar molecular species known to be in the solid state? How are they formed? Do these molecules ever appear in the gas phase?

Molecules in the solid phase are fixed in the structure that makes up the solid, so they cannot rotate but they can vibrate. They show an absorption spectrum in the near infrared in which the rotational substructure is suppressed; it is a pure vibrational spectrum. Species giving rise to an interstellar spectrum of this kind are in an ice mixture on the surface of dust grains. The molecules in such an ice can arise in one of three ways. First, they may start out in the gas, then simply collide and stick to a grain and be absorbed into the surface ice. Carbon monoxide is an example of this kind of species. Second, they may be formed in reactions on a grain surface. A water molecule is a good example: an O atom arrives at a grain surface and is converted into H_2O, retained on the surface and incorporated in the surface ice. Finally, reactions may occur between molecules existing in the ice. These reactions may be stimulated to occur by external energy sources such as ultraviolet starlight or fast particles. Many such reactions have been studied in the laboratory. An example of a new molecule formed from existing ice molecules is acetaldehyde, CH_3CHO. Molecules in the ice can be desorbed into the gas phase if the ices are warmed; this occurs in the process of star formation. The heated molecular-rich gas is in a relatively small region called a hot core, centred on the newly-forming star.

8. Dust has a passive involvement in interstellar chemistry by locking up some elements that would otherwise be present in the gas and available for chemistry. What are its active roles? How important are these roles in the evolution of a galaxy like the Milky Way? Are some environments in the Universe dust-free?

The main passive role for dust is that it contains significant amounts of heavy elements that are removed from the gas and no longer available for processes in the gas. A third of oxygen, more than half the carbon, and nearly all of magnesium, silicon and iron are locked in grains and are unavailable for interstellar chemistry (see Chapter 1). In fundamentally active roles, dust scatters and absorbs starlight and shields the interiors of interstellar clouds from starlight, allowing chemistry to develop there. The energy absorbed by dust is re-radiated in the infrared, allowing the presence of very dense, dark clouds to be mapped. Chemical reactions take place on dust surfaces, principally in the formation of molecular hydrogen without which interstellar chemistry is severely constrained. In denser gas, dust surfaces become coated with molecular ices; an active solid state chemistry is stimulated in these ices by energetic particles or by ultraviolet or X-ray radiation and provides a very rich chemistry of organic molecules, supplementing those produced in the gas phase. Laboratory experiments demonstrating the effectiveness of this solid state chemistry in producing many of the detected species were described in Chapter 8.

9. The list (see the Appendix to Chapter 1) includes carbon species in the form of both rings and chains. Some recent detections are of single-ring PAH molecules, and one 2021 detection is of indene (fused benzene and cyclopentene rings), while the chain $HC_{11}N$ (cyanopenta-acetylene) was also detected in 2021. Is it likely that both types of structure – rings and chains – can be formed in similar ways or are separate mechanisms required?

   Reaction networks of ion-molecule gas-phase reactions such as those that have been described in Chapter 5 have been proposed to account for the formation of both ring and chain molecules. These schemes involved the growth of chains which may then fold up to form rings. It seems probable that these schemes can account for the presence of many of the detected ring species in interstellar and circumstellar space. The schemes are still being evaluated. An alternative approach to account for PAH-type molecules is to consider the reverse process, in which solid carbon dust grains incorporating rings are eroded, so that erosion products – rings and chains – enter the gas phase. This process requires detailed evaluation.

10. In Section 1.1.1, we noted that isotopologues are present and may be relatively abundant in space. How are these isotopologues formed? Why are isotopologues involving minor isotopes often much more abundant than expected?

While the abundance of deuterium relative to hydrogen is small, about 1.6×10^{-5}, the abundance of deuterium-containing molecules in dark, cold interstellar clouds relative to the main isotopologue can be much larger. For example, the relative abundance of HDCO to H_2CO in the Orion molecular cloud is 0.14, a factor of $\sim 10^4$ larger than one might naively expect. Minor isotopologues (*e.g.*, HDCO) are usually formed by reaction of the major isotopologue (*e.g.*, H_2CO) with a deuterium-rich ion such as H_2D^+ in simple ion-molecule reactions such as

$$H_2D^+ + H_2CO \rightarrow H_2DCO^+ + H_2$$

$$H_2DCO^+ + e \rightarrow HDCO + H$$

The ion H_2D^+ is strongly enriched with deuterium because it is formed in the reversible reaction with HD (the main reservoir of deuterium in cold clouds)

$$H_3^+ + HD \leftrightarrow H_2D^+ + H_2$$

in which the products on the right hand side are slightly lower in energy, by an amount equivalent to 178 K (because of a small difference in zero-point energy between H_3^+ and H_2D^+). This means that at temperatures above that value, the forward and back reactions are equally likely, but at lower temperatures (such as those in dark interstellar clouds) only the forward channel is fully open, so that the protonated molecular hydrogen becomes very rich in deuterium in low temperature clouds, a process known as *fractionation*. This strong enhancement in deuterium can then be transmitted to other molecules (such as H_2CO) in low temperature ion-molecule chemistry. Similar mechanisms apply for reactions involving other isotopes. For example, the reaction

$$^{13}C^+ + {}^{12}CO \leftrightarrow {}^{12}C^+ + {}^{13}CO$$

is exothermic in the forward direction by an energy equivalent to 35 K. At temperatures above this value, the reaction may proceed in both directions, but at lower temperatures the reverse reaction is partially inhibited so the abundance of the ^{13}CO isotopologue is enhanced.

10.6 Conclusion

Astrochemistry has made and continues to make huge demands on laboratory and theoretical chemistry. In doing so, astrochemistry has been a stimulus to studying the chemistry that occurs under rather unusual physical conditions. The continuing remarkable advances in chemistry relevant to astronomy enhance enormously our understanding of the Universe and particularly of star and planet formation and the origin of life. It is also thought-provoking, exciting, and – importantly – fun to work in astrochemistry; it is a challenging and rapidly expanding area of science. However, much remains to be done. It would be wrong to give the impression that astrochemistry is fully understood. In this book we have emphasised the main processes by which molecules can be synthesized in interstellar and circumstellar space, but the gas-phase and solid–state pathways to forming some of the larger detected species are unclear and remain the topics of current research in astrochemistry. The relationship between astrochemistry and astrobiology remains to be fully explored.

Subject Index

absorption spectra 41–42
accretionary phase 191
acetylene 80–81, 148
achondrites 194, 211
activation energy 91
active galaxies 20
adenine 213
AGB (Asymptotic Giant Branch)
    stars 135, 137, 138–148
alanine 212
albedo 18
ALMA (Atacama Large Millimetre
    Array) 233–234
aluminium oxide 147
amino acids 211–212, 216–217
aminonitriles 212
ammonia 177
anharmonic oscillator model 39–40
anions, interstellar 66–67
anoxic atmospheres 218
Arrhenius rate law 90, 91–92
associative detachment 66
asteroids 192, 193
astrobiology 186–187, 204–216
astronomical observatories 231–234
astronomical unit ix
asymmetric top molecules 37–38
Asymptotic Giant Branch (AGB) stars
    135, 137, 138–148
Atacama Large Millimetre Array
    (ALMA) 233–234
atmosphere, Earth 198–199, 218–221,
    231–232
atmospheres, planetary 195–200,
    202–203
atomic hydrogen *see* hydrogen (atomic)
Auger effect 107

baryonic matter 110–112
Big Bang 110, 112

bimolecular reactions 86, 96–97
biomarkers 204–205
biosignatures 204–205
black dwarfs 137
bombardment phase 192
brightness temperature 46–47
Burcat, Alexander 105–106

calcium–aluminium-rich inclusions
    (CAIs) 192
carbon atoms 80–82
carbon dioxide, in planetary
    atmospheres 199
carbon insertion reactions 80–82
carbon ions 80–82
carbon monosulfide (CS) 79
carbon monoxide (CO)
    critical number density 44
    dissociative charge exchange
        reactions 55
    first detection in interstellar
        space 3
    formation 78–79
    hydrogenation 86
    ice 176–178, 182
    in circumstellar regions 139
    partial pressure 143
    spectra 36, 38, 40
carbon-rich stellar atmospheres
    138–141, 147
carbonaceous chondrites 195
Cepheid Variable stars 135–136
cgs (centimetre-gram-seconds)
    units xi
chain molecules 240
chemical kinetics 86–94
chemisorption 98, 165–166
chiral molecules 213–214
chondrites 191, 193–194, 195, 211
chondrules 191–192

242